STARGAZING 2013

MONTH-BY-MONTH GUIDE TO THE NORTHERN NIGHT SKY

HEATHER COUPER & NIGEL HENBEST

www.philips-maps.co.uk

HEATHER COUPER and NIGEL HENBEST are internationally recognized writers and broadcasters on astronomy, space and science. They have written more than 30 books and over 1000 articles, and are the founders of an independent TV production company specializing in factual and scientific programming.

Heather is a past President of both the British Astronomical Association and the Society for Popular Astronomy. She is a Fellow of the Royal Astronomical Society, a Fellow of the Institute of Physics and a former Millennium Commissioner, for which she was awarded the CBE in 2007. Nigel has been Astronomy Consultant to *New Scientist* magazine, Editor of the *Journal of the British Astronomical Association* and Media Consultant to the Royal Greenwich Observatory.

Published in Great Britain in 2012
by Philip's,
a division of Octopus Publishing Group Limited
(www.octopusbooks.co.uk)
Endeavour House, 189 Shaftesbury Avenue,
London WC2H 8JY
An Hachette UK Company (www.hachette.co.uk)

TEXT
Heather Couper and Nigel Henbest (pages 6–53)
Robin Scagell (pages 61–64)
Philip's (pages 1–5, 54–60)

ISBN 978–1–84907–235–9

Printed in China

Title page: *The Flame Nebula and Horsehead Nebula (Steve Porter/Galaxy)*

ACKNOWLEDGEMENTS

All star maps by Wil Tirion/Philip's, with extra annotation by Philip's.
Artworks © Philip's.

All photographs courtesy of Galaxy Picture Library:
James Dyson *36;*
Richard Fleet *64;*
Mike Harlow *48;*
Geir T. Øye *24;*
Emma Porter *16;*
Steve Porter *8;*
Gordon Rogers *41;*
Robin Scagell *12, 32, 61–63;*
Peter Shah *44, 52;*
Michael Stecker *28;*
Dave Tyler *20.*

CONTENTS

The sight of diamond-bright stars sparkling against a sky of black velvet is one of life's most glorious experiences. No wonder stargazing is so popular. Learning your way around the night sky requires nothing more than patience, a reasonably clear sky and the 12 star charts included in this book.

Stargazing 2013 is a guide to the sky for every month of the year. Complete beginners will use it as an essential night-time companion, while seasoned amateur astronomers will find the updates invaluable.

THE MONTHLY CHARTS

Each pair of monthly charts shows the views of the heavens looking north and south. They are usable throughout most of Europe – between 40 and 60 degrees north. Only the brightest stars are shown (otherwise we would have had to put 3000 stars on each chart, instead of about 200). This means that we plot stars down to third magnitude, with a few fourth-magnitude stars to complete distinctive patterns. We also show the ecliptic, which is the apparent path of the Sun in the sky.

USING THE STAR CHARTS

To use the charts, begin by locating the north Pole Star – Polaris – by using the stars of the Plough (see March). When you are looking at Polaris you are facing north, with west on your left and east on your right. (West and east are reversed on star charts because they show the view looking up into the sky instead of down towards the ground.) The left-hand chart then shows the view you have to the north. Most of the stars you see will be circumpolar, which means that they are visible all year. The other stars rise in the east and set in the west.

Now turn and face the opposite direction, south. This is the view that changes most during the course of the year. Leo, with its prominent 'sickle' formation, is high in the spring skies. Summer is dominated by the bright trio of Vega, Deneb and Altair. Autumn's familiar marker is the Square of Pegasus, while the winter sky is ruled over by the stars of Orion.

The charts show the sky as it appears in the late evening for each month: the exact times are noted in the caption with the chart. If you are observing in the early morning, you will find that the view is different. As a rule of thumb, if you are observing two hours later than the time suggested in the caption, then the following month's map will more accurately represent the stars on view. So, if you wish to observe at midnight in the middle of February, two hours later than the time suggested in the caption, then the stars will appear as they are on March's chart. When using a chart for the 'wrong' month, however, bear in mind that the planets and Moon will not be shown in their correct positions.

THE MOON, PLANETS AND SPECIAL EVENTS

In addition to the stars visible each month, the charts show the positions of any planets on view in the late evening. Other planets may also be visible that month, but they will not be on the chart if they have already set, or if they do not rise until early morning. Their positions are described in the text, so that you can find them if you are observing at other times.

We have also plotted the path of the Moon. Its position is marked at three-day intervals. The dates when it reaches First Quarter, Full Moon, Last Quarter and New Moon are given in the text. If there is a meteor shower in the month, we mark the position from which the meteors appear to emanate – the *radiant*. More information on observing the planets and other Solar System objects is given on pages 54–57.

Once you have identified the constellations and found the planets, you will want to know more about what's on view. Each month, we explain one object, such as a particularly interesting star or galaxy, in detail. We have also chosen a spectacular image for each month and described how it was captured. All of these pictures were taken by amateurs. We list details and dates of special events, such as meteor showers or eclipses, and give observing tips. Finally, each month we pick a topic related to what's on view, ranging from quasars to star names and the origin of the Universe, and discuss it in more detail. Where possible, all relevant objects are highlighted on the maps.

FURTHER INFORMATION

The year's star charts form the heart of the book, providing material for many enjoyable observing sessions. For background information turn to pages 54–57, where diagrams help to explain, among other things, the movement of the planets and why we see eclipses.

Although there is plenty to see with the naked eye, many observers use binoculars or telescopes, and some choose to record their observations using cameras, CCDs or webcams. For a round-up of what's new in observing technology, go to pages 61–64, where equipment expert Robin Scagell shares his knowledge of planispheres, computer software and apps.

If you have already invested in binoculars or a telescope, then you can explore the deep sky – nebulae (starbirth sites), star clusters and galaxies. On pages 58–60 we list recommended deep-sky objects, constellation by constellation. Use the appropriate month's maps to see which constellations are on view, and then choose your targets. The table of 'limiting magnitude' (page 58) will help you to decide if a particular object is visible with your equipment.

Happy stargazing!

If ever there was a time to see A-list stars strutting their stuff, it's this month. The constellations of winter are so striking that there's no better time to start finding your way around the sky. And this year, they're joined by an interloper – the giant planet **Jupiter**. It joins the dazzling denizens of **Orion**, **Taurus**, **Gemini** and **Canis Major** to make up a scintillating celestial tableau.

▼ The sky at 10 pm in mid-January, with Moon positions at three-day intervals either side of Full Moon. The star positions are also correct for 11 pm at

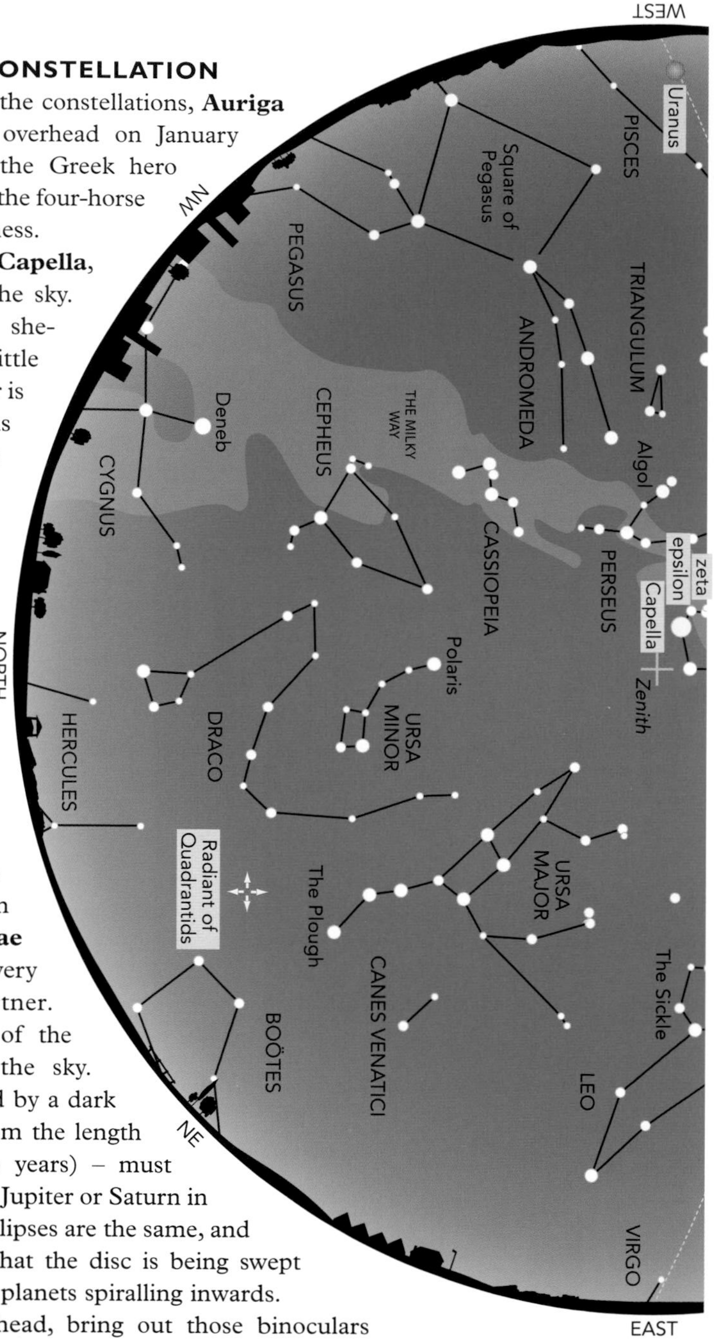

JANUARY'S CONSTELLATION

One of the most ancient of the constellations, **Auriga** (the Charioteer), sparkles overhead on January nights. It is named after the Greek hero Erichthoneus, who invented the four-horse chariot to combat his lameness.

Auriga is dominated by **Capella**, the sixth-brightest star in the sky. Its name means 'the little she-goat', but there's nothing little about Capella. The giant star is over 150 times more luminous than our Sun, and it also has a yellow companion.

Capella, the goat, marks the Charioteer's shoulder, and – to her right – is a tiny triangle of stars, the lower pair nicknamed 'the kids'. Two of the stars in the trio are variable stars – but not because of any intrinsic instability. They're 'eclipsing binaries': stars which change in brightness because a companion star passes in front of them. **Zeta Aurigae** is an orange star eclipsed every 972 days by a blue partner. **Epsilon Aurigae** is one of the weirdest star systems in the sky. Every 27 years, it is eclipsed by a dark disc of material which – from the length of the eclipses (nearly two years) – must extend to that of the orbit of Jupiter or Saturn in our Solar System. No two eclipses are the same, and there are tantalizing hints that the disc is being swept clean by proto-Jupiter-sized planets spiralling inwards.

And when Auriga's overhead, bring out those binoculars

the beginning of January, and 9 pm at the end of the month. The planets move slightly relative to the stars during the month.

(or better still, a small telescope) – for within the 'body' of the Charioteer are three very pretty star clusters: **M36, M37** and **M38**.

PLANETS ON VIEW

Jupiter opens the year with a flourish, soaring high in the south in the evening sky and setting around 4.30 am. At magnitude −2.5, the giant planet outshines any of the stars. Jupiter lies in Taurus, in the fringes of the Hyades star cluster – between Aldebaran and the Pleiades. With binoculars or a small telescope, look for Jupiter's four largest moons: you may see more than you expect, as the planet passes various background stars roughly equal in brightness.

The evening sky holds three more difficult planet challenges. **Mars** skulks low in the dusk sky, in Capricornus: shining at magnitude +1.2, the Red Planet sets at 6 pm. It's moving towards **Neptune** (magnitude +8.0), which lies in Aquarius and sets around 7 pm. Both are lost in twilight by the end of January. **Uranus** – on the edge of naked-eye visibility at magnitude +5.9 – lies in Pisces and sets at 11 pm.

Saturn is rising in the south-east at 2 am. The ringworld shines at magnitude +0.9 on the borders of Virgo and Libra.

Just before sunrise, you can catch **Venus** (magnitude −3.8) low in the south-east. At the beginning of the month, it's rising one-and-a-half hours before the Sun, but by the end of January it has sunk from sight in the Sun's glare.

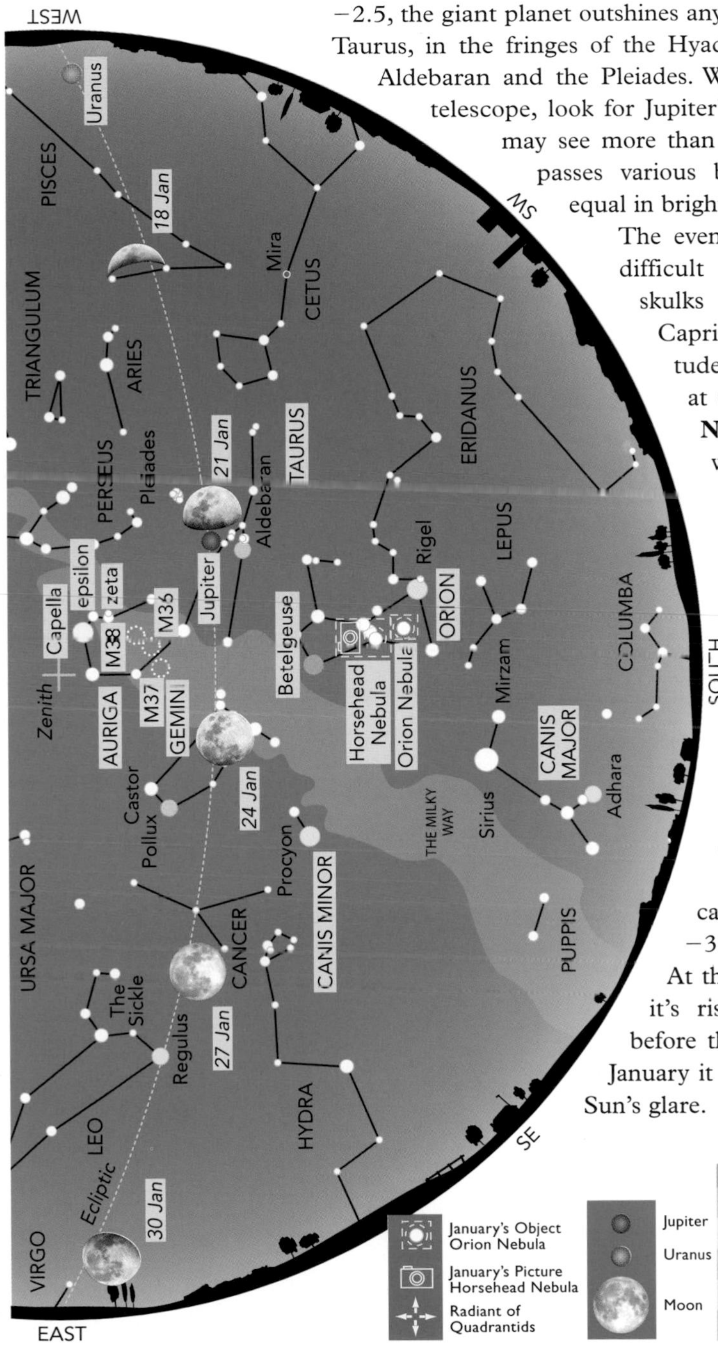

January's Object Orion Nebula
January's Picture Horsehead Nebula
Radiant of Quadrantids
Jupiter
Uranus
Moon

MOON		
Date	Time	Phase
5	3.58 am	Last Quarter
11	7.44 pm	New Moon
18	11.45 pm	First Quarter
27	4.38 am	Full Moon

Mercury is too close to the Sun to be seen this month.

MOON

The Moon lies near Regulus on 1 January. On the morning of 6 January you'll find the Moon below Spica, and it passes underneath Saturn in the morning of 7 January. Before dawn on 10 January the crescent Moon lies near Venus (see Special Events), and the waxing crescent Moon moves above Mars on 13 January. The Moon passes Jupiter and Aldebaran on 21/22 January (see Special Events). The Moon is back with Regulus on 28 January.

SPECIAL EVENTS

On **2 January**, at 0.59 am, the Earth is at perihelion, its closest point to the Sun – a 'mere' 147 million kilometres away.

The maximum of the Quadrantid meteor shower occurs on the night of **3/4 January**. These shooting stars are tiny particles of dust shed by the old comet 2003 EH_1, burning up as they enter the Earth's atmosphere. The show will be spoilt after 10 pm, when the Moon rises.

On **10 January**, the slender crescent Moon makes a striking pair with Venus in the morning sky, around 7 am.

There's a lovely sight on the night of **21/22 January**, when the Moon passes just a degree below giant planet Jupiter, with Aldebaran to the left.

Viewing tip

It may sound obvious, but if you want to stargaze at this most glorious time of year, dress up warmly! Lots of layers are better than a heavy coat as they trap air next to your skin – and heavy-soled boots stop the frost creeping up your legs. It may sound anorakish, but a woolly hat really does stop one-third of your body's heat escaping through the top of your head. And – alas – no hipflask of whisky. Alcohol constricts the veins and makes you feel even colder.

JANUARY'S OBJECT

Below **Orion's** Belt lies a fuzzy patch – easily visible to the unaided eye in dark skies. Through binoculars, or a small telescope, the patch looks like a small cloud in space. It is a cloud – but at 24 light years across, it's hardly petite. Only the distance of the **Orion Nebula** – 1300 light years – diminishes it. Yet it is the nearest region to Earth where heavyweight stars are being born: this 'star factory' contains at least 150 fledgling stars, which have condensed out of dark dust and gas clouds around the nebula. The most prominent are the four stars making up 'the Trapezium' cluster.

JANUARY'S PICTURE

The celestial chess-piece of the Universe: Orion's **Horsehead Nebula**. This dark cloud of dust and gas is poised to create a new generation of stars. To the bottom left is the Flame Nebula. This star-factory has already begun to create its offspring, which illuminate its swirling fronds of gas. The Flame Nebula – which lies about 1000 light years away – is just one of many star-forming regions in Orion.

JANUARY'S TOPIC
Return to the Moon

The starting-gun fires this year for the start of a new Space Race. A Chinese spacecraft due to be launched this month will make the first soft-landing on the Moon in 37 years. Chang'e 3 follows two Chinese spacecraft that have already gone into orbit around our celestial companion. Its robotic explorer will send back pictures and data from a lava plain picturesquely called Sinus Iridum – the Bay of Rainbows.

Later in the year, NASA is launching the LADEE (Lunar Atmosphere and Dust Environment Explorer) mission, to measure the Moon's evanescent cloak of gas and dust, before it's disrupted by future manned missions. Yes, we know the old joke that the problem with pubs on the Moon is they've no atmosphere – but the Apollo astronauts did find traces of dust and gas, blasted out of the lunar surface by micro-meteorites or the Sun's radiation.

Next year, we should see an Indian Moon-lander and a Russian orbiter dropping penetrators deep into the lunar surface. After that, the action really hots up, as 20 international teams of private competitors fight for the Google Lunar X-Prize: the winner must land a rover on the Moon that travels 500 metres and sends back high-definition video – all before the end of 2015.

◀ *The Flame Nebula (left) and Horsehead Nebula (centre) in Orion, photographed by Steve Porter using a William Optics 132 mm refractor and modified Canon 5D camera. The total exposure time was 5¼ hours, in 63 five-minute exposures. The observatory is on Bardsey Island, off the North Wales coast.*

The winter star-patterns are starting to drift towards the west, setting earlier – a sure sign that spring is on the way. The constantly-changing pageant of constellations in the sky is proof that we live on a cosmic carousel, orbiting the Sun. Imagine it: you're in the fairground, circling the merry-go-round on your horse, and looking out around you. At times you spot the ghost train; sometimes you see the roller-coaster; and then you swing past the candy-floss stall. So it is with the sky – and the constellations – as we circle our local star. That's why we get to see different stars in different seasons. Plus – this month – the elusive planet **Mercury** puts in its best appearance in our evening skies.

▼ *The sky at 10 pm in mid-February, with Moon positions at three-day intervals either side of Full Moon. The star positions are also correct for 11 pm at*

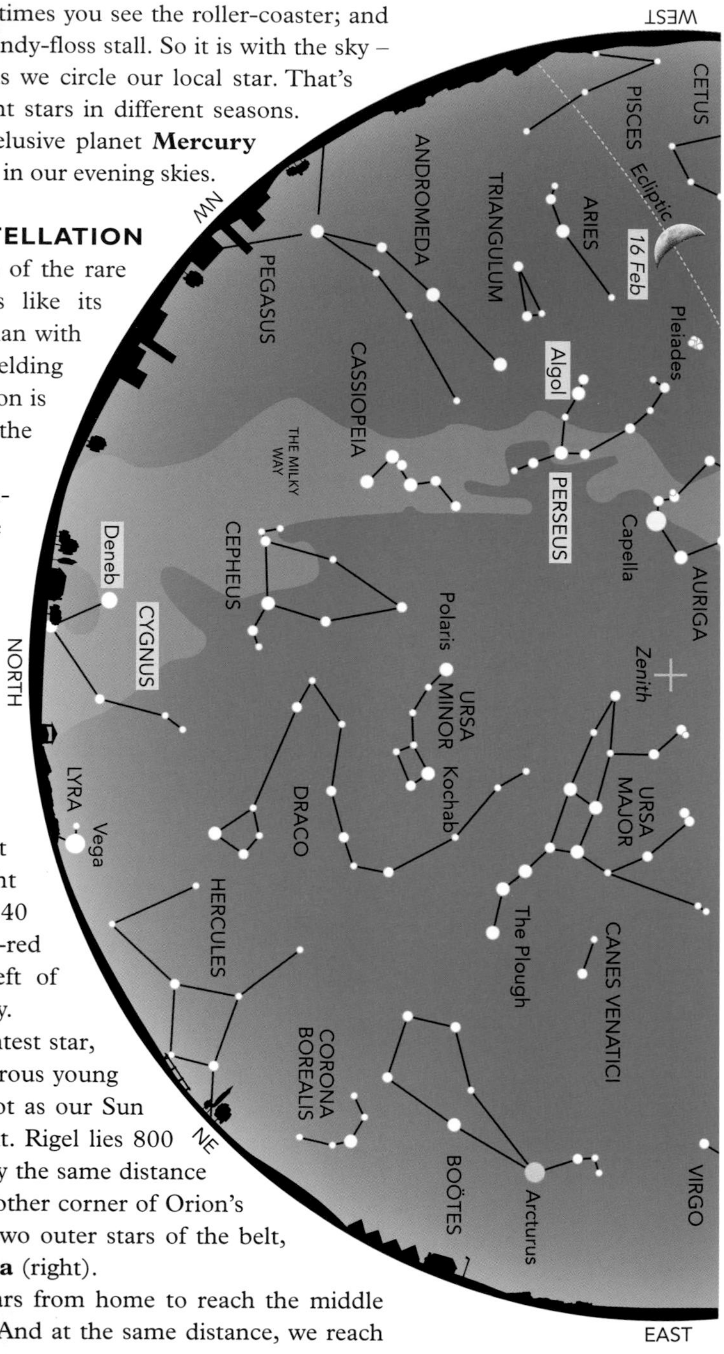

FEBRUARY'S CONSTELLATION

Spectacular **Orion** is one of the rare star-groupings that looks like its namesake – a giant of a man with a sword below his belt, wielding a club above his head. Orion is fabled in mythology as the ultimate hunter.

The constellation contains one-tenth of the brightest stars in the sky: its seven main stars all lie in the 'top 70' of brilliant stars. Despite its distinctive shape, most of these stars are not closely associated with each other – they simply line up, one behind the other.

Closest is the star that forms the hunter's right shoulder, **Bellatrix**, at 240 light years. Next is blood-red **Betelgeuse** at the top left of Orion, 640 light years away.

The constellation's brightest star, blue-white **Rigel**, is a vigorous young star more than twice as hot as our Sun and 50,000 times as bright. Rigel lies 800 light years from us, roughly the same distance as the star that marks the other corner of Orion's tunic – **Saiph** – and the two outer stars of the belt, **Alnitak** (left) and **Mintaka** (right).

We travel 1300 light years from home to reach the middle star of the belt, **Alnilam**. And at the same distance, we reach

the beginning of February, and 9 pm at the end of the month. The planets move slightly relative to the stars during the month.

the stars of the 'sword' hanging below the belt – the lair of the great **Orion Nebula** (see January's Object).

PLANETS ON VIEW

Mercury is putting on its best evening performance of the year, reaching its greatest elongation from the Sun on 16 February. Look for it very low in the west around 6.30 pm in the period 10–21 February, during which its brightness declines from magnitude −0.9 to +0.3.

Uranus, at magnitude +5.9 in Pisces, sets at 9.45 pm at the beginning of February; by the end of the month, it has disappeared into the twilight glow.

But it's the two biggest planets that are centre-stage this month. First on the scene is **Jupiter,** at magnitude −2.3, blazing in the evening skies among the stars of Taurus: Aldebaran lies to its left and the Pleiades to the right. Jupiter sets around 2.30 am.

Saturn rises at midnight in the south-east. Lying in Libra, the ringworld shines yellow-white at magnitude +0.7.

Venus, Mars and **Neptune** are lost in the Sun's glare this month.

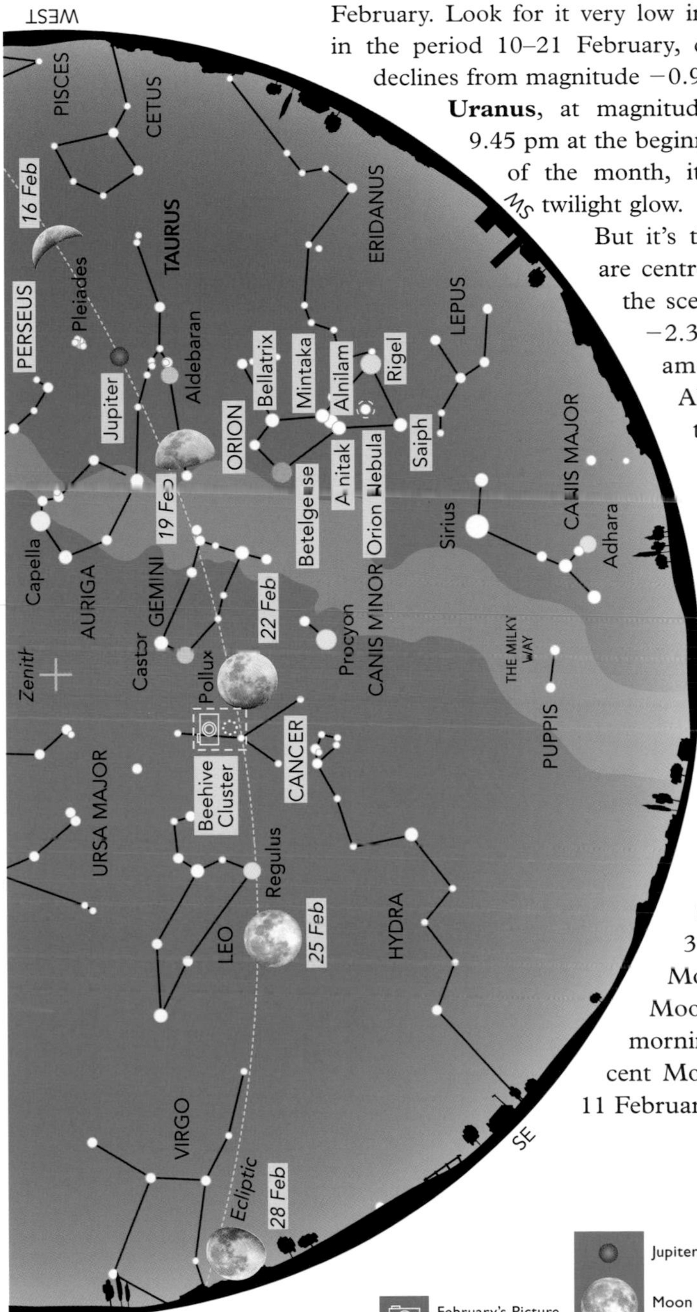

MOON

As the Moon rises at midnight on 1/2 February, it's less than a degree below Spica. On the morning of 3 February, the Last Quarter Moon passes below Saturn. The Moon lies above Antares in the morning of 5 February. The crescent Moon passes near Mercury on 11 February (see Special Events). You'll

February's Picture Beehive Cluster

Jupiter

Moon

MOON		
Date	Time	Phase
3	1.56 pm	Last Quarter
10	7.20 am	New Moon
17	8.31 pm	First Quarter
25	8.26 pm	Full Moon

find the First Quarter Moon near Jupiter on 17 and 18 February, with Aldebaran to the left. The Moon lies below Regulus on 24 February, and passes close to Spica on the night of 28 February/1 March.

SPECIAL EVENTS

Just after sunset on **11 February**, the narrow crescent Moon pairs up with Mercury: the innermost planet lies to the lower left of the Moon.

At 7.26 pm on **15 February**, asteroid 2012 DA_{14} skims past the Earth at a distance of only 22,000 km – within the orbits of TV broadcasting satellites. Some 45 metres across, it will shine at magnitude +7, and should be visible in binoculars.

FEBRUARY'S OBJECT

Look out for hard-to-spot **Mercury** this month. Rumour has it that the architect of our Solar System – Nicolaus Copernicus – never saw the tiny world because of mists rising from the nearby River Vistula in Poland. Being the closest planet to the Sun, it seldom strays far from the glare of our local star.

The pioneering space probe Mariner 10 sent back only brief images as it swung past the diminutive, cratered planet in 1974. But on 18 March 2011, everything changed: that's when the NASA probe Messenger went into orbit around Mercury.

Messenger (the convoluted acronym stands for ME-rcury S-urface S-pace EN-vironment GE-ochemistry and R-anging mission) celebrates the belief that, in mythology, fleet-footed Mercury was messenger to the gods. Messenger made three flybys of Mercury, to slow down the probe.

Now in Mercury-orbit, Messenger's instruments are scanning the planet's surface and scrutinizing its composi-

▼ Comet NEAT (C/2001 Q4) photographed by Robin Scagell on 16 May 2004 near the Beehive Cluster. A Canon 10D camera was used at ISO 200 with a 135 mm lens. The total exposure time over four frames was 4 minutes 10 seconds.

tion – offering clues as to the origins of this mysterious body. Researchers suspect that the tiny world may have been born further out from the Sun and later spiralled in.

Messenger is also exploring Mercury's internal workings – and it's indicating that our innermost world has a core made of molten iron. Plus: it has discovered water in Mercury's outer atmosphere, and evidence for past volcanic activity.

Viewing tip

Don't think that you need a telescope to bring the heavens closer. Binoculars are excellent – and you can fling them into the back of the car at the last minute. But when you buy binoculars, make sure that you get those with the biggest lenses, coupled with a modest magnification. Binoculars are described, for instance, as being '7 × 50' – meaning that the magnification is seven times, and that the diameter of the lenses is 50 mm. These are ideal for astronomy – they have good light grasp, and the low magnification means that they don't exaggerate the wobbles of your arms too much. It's always best to rest your binoculars on a wall or a fence to steady the image. Some amateurs are the lucky owners of huge binoculars – say, 20 × 70 – with which you can see the rings of Saturn (being so large, these binoculars need a special mounting). But above all, never buy binoculars with small lenses that promise huge magnifications – they're a total waste of money.

FEBRUARY'S PICTURE

The **Beehive** star cluster in **Cancer** was wonderfully chronicled by the ancient Chinese as 'the exhalation of piled-up corpses'. In a faint and uninteresting constellation, the cluster is the icing on the cake – and it's just visible to the unaided eye. The Romans called it Praesepe, 'the manger'. With his pioneering telescope, Galileo found this patch is actually a cluster of stars: it contains about 1000 stars and lies about 600 light years away. From its motion through space, astronomers suspect that the Beehive was born alongside the Hyades star cluster in Taurus.

FEBRUARY'S TOPIC
Star names

Why do the brightest stars have such strange names? The reason is that they date from antiquity, and have passed on down generations ever since. The original western star names – like the original constellations – were probably Babylonian or Chaldean, but few of these survive. The Greeks took up the baton after that, and the name of the star **Antares** (see June's Object) is a direct result. It means 'rival of Ares' because its red colour rivals that of the planet Mars (Ares in Greek).

But the Arabs were largely responsible for the star names we have inherited today. Working in the so-called 'Dark Ages' between the 6th and 10th centuries AD, they took over the naming of the sky – hence the number of stars beginning with the letters 'al' (Arabic for 'the'). **Algol**, in the constellation **Perseus**, means 'the demon' – possibly because the Arabs noticed that its brightness seems to 'wink' every few days. **Deneb**, in **Cygnus**, also has Arabic roots – it means 'the tail' (of the flying bird).

The most famous star in the sky has to be **Betelgeuse** – known to generations of school kids as 'Beetlejuice'. It was gloriously interpreted to mean 'the armpit of the sacred one'. But the 'B' in Betelgeuse turned out to be a mis-transliteration – and so we're none the wiser as to how our distant ancestors really identified this fiery red star.

This month, the nights become shorter than the days as we hit the Vernal Equinox – on 20 March, spring is 'official'. That's the date when the Sun climbs up over the equator to shed its rays over the northern hemisphere. Because of the Earth's inclination of 23.5° to its orbital path around the Sun, the North Pole points away from our local star between September and March, causing the long nights of autumn and winter. Come the northern spring, Earth's axial tilt means that the Sun favours the north – and we can look forward to the long, warm days of summer. Even better, the clocks 'spring forward' on 31 March, making the evenings lighter.

▼ The sky at 10 pm in mid-March, with Moon positions at three-day intervals either side of Full Moon. The star positions are also correct for 11 pm at

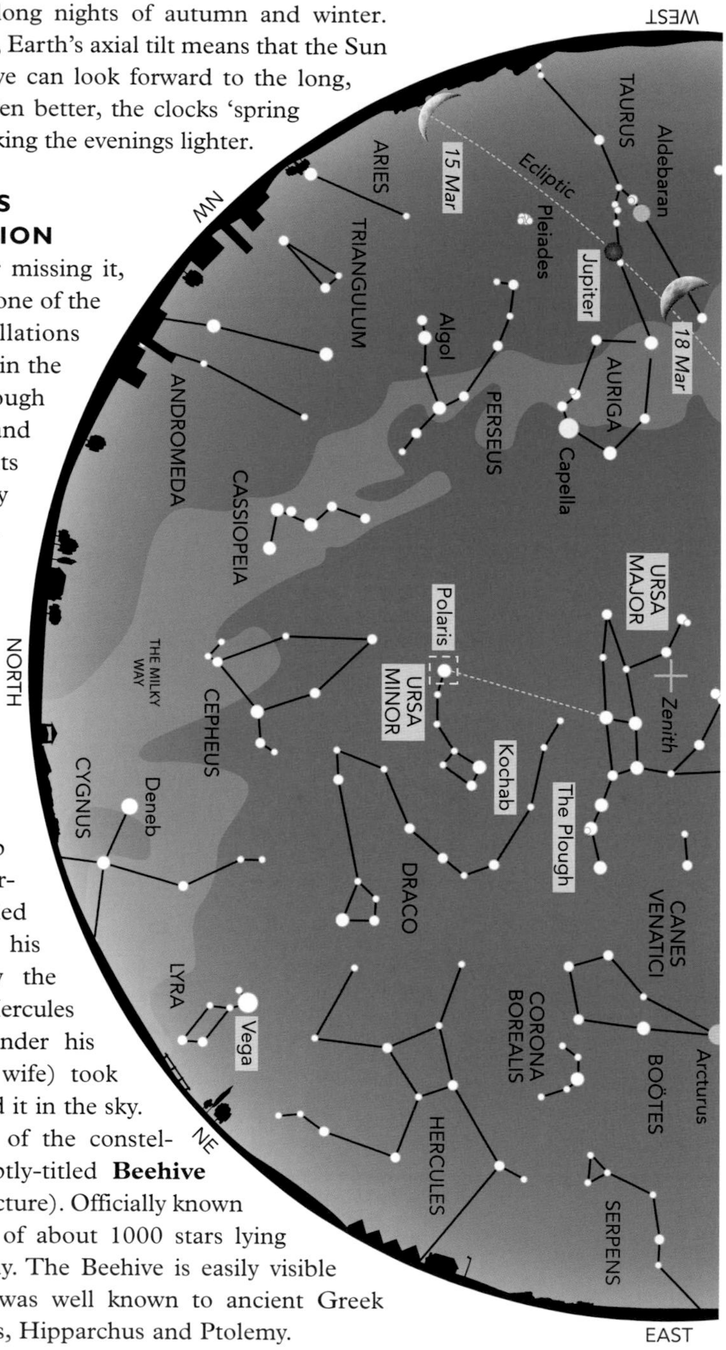

MARCH'S CONSTELLATION

You could be forgiven for missing it, because **Cancer** is hardly one of the most spectacular constellations in the sky. Although it lies in the Zodiac – the band through which the Sun, Moon and planets appear to move – its stars are so faint that city lights completely drown them out. If you do have dark skies, look between the **Sickle** of **Leo** and the twin stars **Castor** and **Pollux** in **Gemini**, and you'll locate the slender little constellation.

According to legend, Cancer is named after the crab that attempted to nip Hercules during his altercation with the multi-headed monster Hydra – one of his '12 labours' ordered by the Oracle at Delphi. Alas, Hercules crushed the crustacean under his foot. But Juno (Jupiter's wife) took pity on the crab and placed it in the sky.

However, at the centre of the constellation is a gem – the aptly-titled **Beehive Cluster** (see February's Picture). Officially known as Praesepe, it's a swarm of about 1000 stars lying nearly 600 light years away. The Beehive is easily visible to the unaided eye, and was well known to ancient Greek astronomers such as Aratos, Hipparchus and Ptolemy.

the beginning of March, and 10 pm at the end of the month (after BST begins). The planets move slightly relative to the stars during the month.

PLANETS ON VIEW

Jupiter and Saturn alone strut their stuff in our skies this month. **Jupiter** is brilliant in Taurus, dominating the evening skies to the west until it sets about 1 am. At magnitude −2.1, the giant world outshines even the magnificent stars of Orion and his entourage. With binoculars or a small telescope, look for its four brightest moons.

Around 10 pm, Jupiter is joined by fellow giant **Saturn**, rising in the east in Libra. At magnitude +0.6, the ringed planet is ten times fainter than Jupiter, but it's still prominent in a region of faint stars.

Venus, **Mars**, **Uranus** and **Neptune** are too close to the Sun to be seen in March. **Mercury**, too, is lost in the twilight glow as seen from the UK, though it's at greatest elongation west on 31 March.

MOON

On the mornings of 2 and 3 March, you'll find the Moon not far from Saturn, and on 4 and 5 March it's near Antares. The Moon passes close to Jupiter on 17/18 March (see Special Events). It lies below Regulus on 24 March, while the Moon is near Spica on 28 March, and passes below Saturn on 29 March.

SPECIAL EVENTS

On the night of **17/18 March**, the waxing Moon passes Jupiter, with Aldebaran and the Hyades to their left.

The Vernal Equinox, on **20 March** at 11.02 am, marks the begin-

March's Object
Polaris

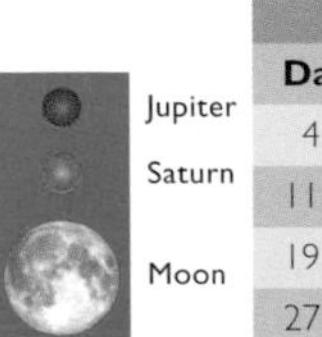

MOON		
Date	Time	Phase
4	9.53 pm	Last Quarter
11	7.51 pm	New Moon
19	5.27 pm	First Quarter
27	9.27 am	Full Moon

ning of spring, as the Sun moves up to shine over the northern hemisphere.

31 March, 1.00 am: British Summer Time starts – don't forget to put your clocks forward (the mnemonic is 'Spring forward, Fall back').

▲ *North polar star-trails photographed by Emma Porter from Llanbister, Powys. A fixed Canon 5D camera was used to take 400 30-second shots, which were later stacked in Photoshop to give the equivalent of a continuous 3 hours 20 minutes' exposure time, though without the build-up of sky background light. A meteor trail is visible at bottom left.*

MARCH'S OBJECT

The Pole Star – **Polaris** – is a surprisingly shy animal, coming in at the modest magnitude of +2.0. You can find it by following the two end stars of the **Plough** (see star chart) in **Ursa Major**, the Great Bear. Polaris lies at the end of the tail of the Lesser Bear (**Ursa Minor**), and it pulsates in size, making its brightness vary slightly over a period of 4 days. But its importance throughout recent history centres on the fact that Earth's North Pole points towards Polaris, so we spin 'underneath' it. It remains almost stationary in the sky, and acts as a fixed point for both astronomy and navigation. However, over a 26,000-year period, the Earth's axis swings around like an old-fashioned spinning top – a phenomenon called precession

Viewing tip
This is the time of year to tie down your compass points – the directions of north, south, east and west as seen from your observing site. North is easy – just latch on to Polaris, the Pole Star (see this month's Object). And at noon, the Sun is always in the south. But the useful extra in March is that we hit the Spring (Vernal) Equinox, when the Sun rises due east, and sets due west. So remember those positions relative to a tree or house around your horizon.

– so our 'pole stars' change with time. Polaris will be nearest to the 'above pole' position in 2100, before the Earth wobbles off. Famous pole stars of the past include Kochab in Ursa Minor, which presided over the skies during the Trojan Wars of 1184 BC. In 12,000 years' time, brilliant **Vega**, in Lyra, will be our pole star.

MARCH'S PICTURE

The glory of our swirling heavens, captured in time-lapse as we spin under the Pole Star. **Star-trails** reveal the rotation of the Earth, and this photograph shows the tracks of stars crossing the sky as the Earth turns. You can see from this image that Polaris (the bright object in the 'bull's-eye') is not exactly above the Earth's North Pole – the star's motion subtends a small arc.

MARCH'S TOPIC
Quasars

The constellation of Virgo is rising this month – and with it, one of the most mysterious and exciting objects in the cosmos: the quasar 3C 273. Discovered by radio astronomers, 'quasars' looked like stars – but they emitted powerful radio waves. Astronomers nicknamed them 'quasi-stellar radio sources' (hence the convoluted acronym!).

Exactly 50 years ago, a young Dutch astronomer, Maarten Schmidt, had the bright idea to take a spectrum of the object with the Hale Telescope on Palomar Mountain (then the largest in the world). When he analyzed the light patterns, he discovered that 3C 273 was rushing away from us at a colossal speed – a result of the expansion of the Universe. This meant one thing: the quasar had to be incredibly distant, and awesomely powerful.

Schmidt's redshift put 3C 273 at a distance of 2.5 billion light years. And this was just the beginning. Of the 200,000 quasars known to exist today, the furthest lies a staggering 30 billion light years away!

Quasars are baby galaxies in their disruptive birth throes. At their core lurks a supermassive black hole, hell-bent on gobbling stars and gas – and their foetal burps cause massive explosions in the Universe.

3C 273 is one of our nearest quasars. And if you have a medium-sized telescope, now (and next month) is the time to home in on the voracious beast. It changes in brightness depending on its eating patterns, but it's now averaging magnitude +12.7.

The springtime skies are dominated by the ancient constellations of **Leo** and **Virgo**. Leo does indeed look like a recumbent lion, but it's hard to envisage Virgo as anything other than a vast 'Y' in the sky!

To the left of Virgo you will find the planet **Saturn**, closest to the Earth this month. 'Close', however, is relative – the ringworld is over a billion kilometres distant.

▼ *The sky at 11 pm in mid-April, with Moon positions at three-day intervals either side of Full Moon. The star positions are also correct for midnight at the beginning of*

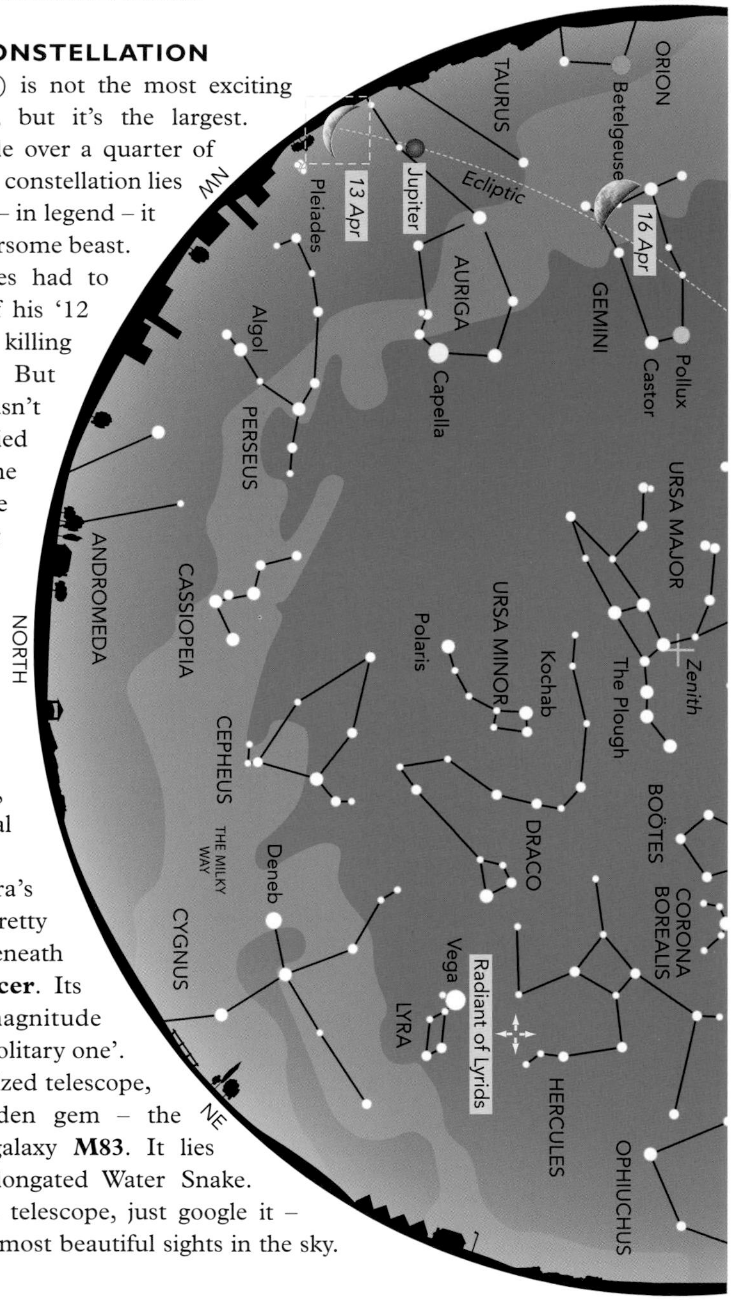

APRIL'S CONSTELLATION

Hydra (the Water Snake) is not the most exciting constellation in the sky, but it's the largest. Hydra's faint stars straggle over a quarter of the sky (100 degrees). The constellation lies south of mighty **Leo**, and – in legend – it is meant to represent a fearsome beast.

The superhero Hercules had to slay the Hydra as one of his '12 labours' – a penance for killing his wife and children. But despatching the Hydra wasn't easy. Even grown men died of fright when they saw the monster. And it had the irksome habit of growing numerous heads – if one was chopped off, three would grow back!

Not to be daunted, Hercules hacked away the extra heads, cauterizing the stumps with burning branches. Finally, he severed the last immortal head with his sword.

In the heavens, Hydra's head is no more than a pretty grouping of faint stars beneath the constellation of **Cancer**. Its main star is second-magnitude **Alphard** – meaning 'the solitary one'.

If you have a medium-sized telescope, search out Hydra's hidden gem – the glorious face-on spiral galaxy **M83**. It lies under the tail of the elongated Water Snake. And if you haven't got a telescope, just google it – this star-city is one of the most beautiful sights in the sky.

April, and 10 pm at the end of the month. The planets move slightly relative to the stars during the month.

PLANETS ON VIEW

Jupiter is king of the evening sky. At magnitude −1.9 in Taurus, it is setting in the west around midnight. Jupiter lies above Aldebaran and the Hyades, with the Pleiades to the lower right.

The planet of the month, though, has to be Saturn. A brilliant jewel in the dull constellation of Libra, Saturn (magnitude +0.3) is at opposition on 28 April and is visible all night long. Grab the chance to look through a small telescope, if you can, to see its glorious rings and its biggest moon, cloud-shrouded Titan.

Mercury, **Venus**, **Mars**, **Uranus** and **Neptune** are all too close to the Sun to be visible in April.

MOON

On the morning of 1 April, the Moon lies near Antares. The crescent Moon passes close to Jupiter on 14 April (see Special Events). The Moon lies beneath Regulus on 20 April. On the night of 24/25 April, the Moon passes less than a degree from Spica. The Full Moon is just below Saturn on 25 April. The Moon lies above Antares in the morning of 28 April.

SPECIAL EVENTS

On **14 April**, we are treated to the lovely sight of Jupiter close up and personal to the crescent Moon, with Aldebaran and the Hyades lying below.

21/22 April: It's the maximum of the **Lyrid** meteor shower, which – by perspective – appears to emanate

MOON		
Date	Time	Phase
3	5.36 am	Last Quarter
10	10.35 am	New Moon
18	1.31 pm	First Quarter
25	8.57 pm	Full Moon

▲ *Saturn, photographed on 15 March 2012 through a 355 mm Schmidt-Cassegrain telescope from Flackwell Heath in Buckinghamshire by Dave Tyler. A Point Grey Flea3 camera was used with separate red, green and blue filters to assemble the colour image. The planet's northern hemisphere is on show.*

from the constellation of Lyra. The shower consists of particles from Comet Thatcher. Unfortunately, the display is spoilt by bright moonlight this year.

On **25 April**, we're treated to the best eclipse visible from the UK in 2013. This partial lunar eclipse reaches its maximum at 9.08 pm, when you'll see a tiny nick taken out of the Moon's top edge. In fact, only 1% of the Moon is covered by the Earth's shadow – which shows just what an appalling year this is for eclipses (see this month's Topic)!

APRIL'S OBJECT

The **Moon** is our nearest celestial companion, lying a mere 384,400 kilometres away. It took the Apollo astronauts only three days to reach it! And at 3476 kilometres across, it's so large when compared to Earth that – from space – the system would look like a double planet.

But the Moon couldn't be more different from our verdant Earth. Almost bereft of an atmosphere, it has been exposed to bombardment by meteorites and asteroids throughout its life. Even with the unaided eye, you can see the evidence. The 'face' of the 'Man in the Moon' consists of huge craters created by asteroid hits 3.8 billion years ago.

Through binoculars or a telescope, the surface of the Moon looks amazing – as if you're flying over it. But don't observe our satellite when it's Full: the light is flat and swamps its features. It's best to roam the Moon when it's a crescent or half-lit, and see the sideways-on shadows highlighting its dramatic relief.

Viewing tip

When you first go out to observe, you may be disappointed at how few stars you can see in the sky. But wait for around 20 minutes and you'll be amazed at how your night vision improves. One reason for this 'dark adaption' is that the pupil of your eye gets larger to make the best of the darkness. More importantly, in dark conditions the retina of your eye builds up much bigger reserves of rhodopsin, the chemical that responds to light.

APRIL'S PICTURE

The most sensational ringworld of our Solar System – **Saturn** (see May's Topic). Its rings, composed of chunks of ice, would stretch nearly all the way from the Earth to the Moon. Saturn is made almost entirely of gas, and is incredibly light. In fact, if you could find an ocean big enough in which to put Saturn, it would float. The gas giant is circled by at least 62 moons.

APRIL'S TOPIC
Unusual eclipses

The year 2013 sees three eclipses of the Moon and two of the Sun – but none of them will be conventional, nor spectacular as seen from the UK. The lunar eclipses (in April, May and October) will be either partial or penumbral.

Eclipses of the Moon take place when our satellite skulks in the Earth's shadow – that's when it, the Earth and the Sun are in line. These dimmings don't happen every month, because the Moon's orbit is angled to that of our planet, and we usually see a Full Moon, with the Sun's rays hitting it face-on.

But the Moon can stray into the darkest part of the shadow – the umbra – and that's when we get a total lunar eclipse. If it doesn't make it fully into the umbra, we're left with a partial eclipse – which is what we'll see this month. May and October offer even less spectacular penumbral eclipses. On these occasions, the Moon glides into the periphery of the Earth's shadow, where some sunlight can still get through – so the Moon will look only slightly washed-out.

May's solar eclipse won't be total, either, because the Moon – which overlaps the Sun during an eclipse – is at its far point, and won't be able to completely cover our local star. So we see an annular eclipse: the Moon blocking out most of the Sun's rays, but surrounded by an annulus of the Sun's brilliant light.

November's solar eclipse is a very weird one. It's a 'hybrid eclipse' – quite a rare entity! The Moon is on the verge of covering the entire Sun – but not quite. Which means that, at the beginning and end of the track, the eclipse is annular. But in the middle of the eclipse path, in central Africa, the end of the Moon's shadow just touches the Earth's surface, providing a brief but wonderful view of a total solar eclipse.

A sure sign that warmer days are here is the appearance of **Arcturus** – a distinctly orange-coloured star that lords it over a huge area of sky devoid of other bright stars. This red giant is the brightest star in the constellation of **Boötes** (the Herdsman), who shepherds the two bears – **Ursa Major** and **Ursa Minor** – through the heavens. Summer is on the way!

▼ The sky at 11 pm in mid-May, with Moon positions at three-day intervals either side of Full Moon. The star positions are also correct for midnight at the beginning of

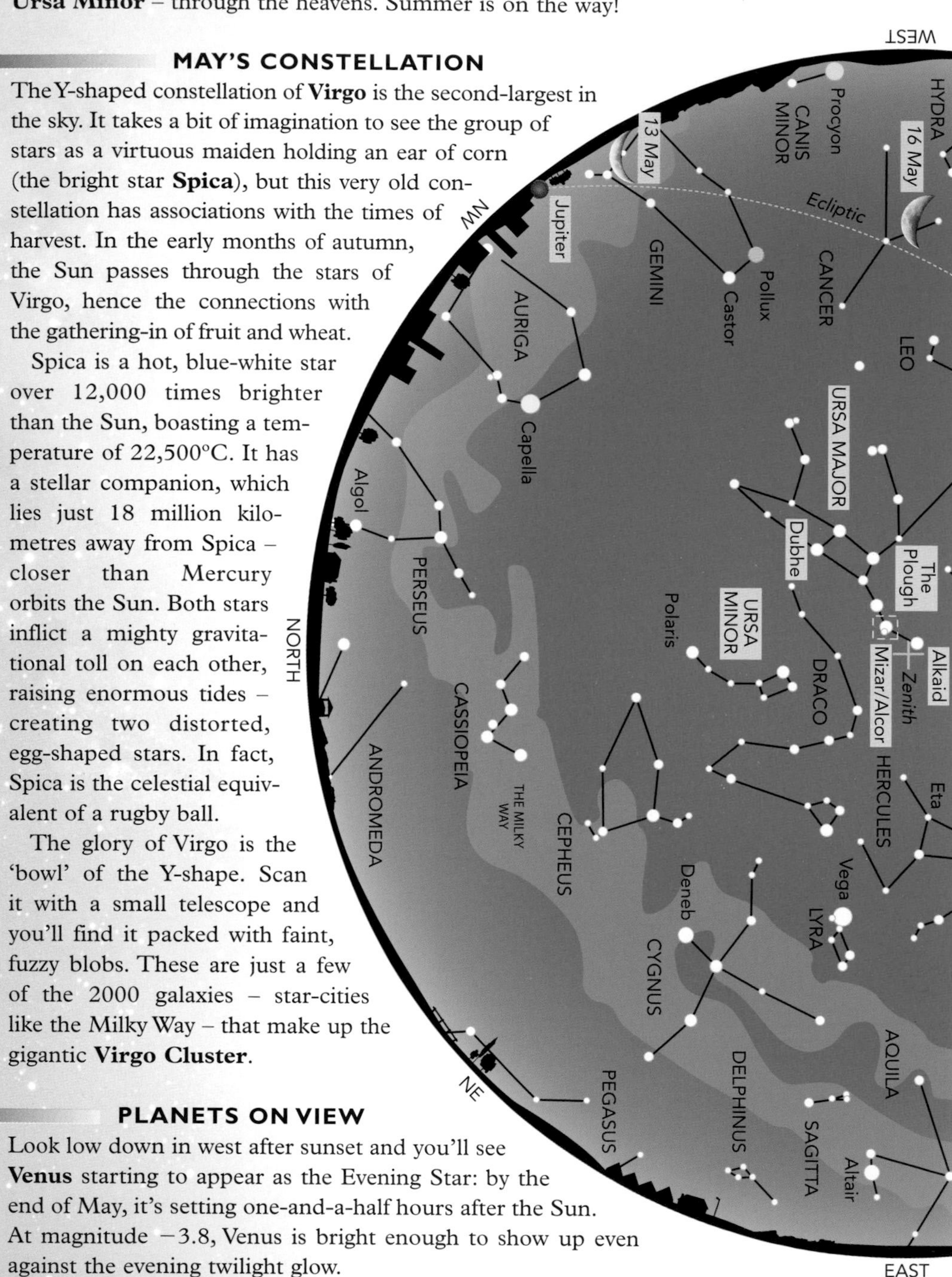

MAY'S CONSTELLATION

The Y-shaped constellation of **Virgo** is the second-largest in the sky. It takes a bit of imagination to see the group of stars as a virtuous maiden holding an ear of corn (the bright star **Spica**), but this very old constellation has associations with the times of harvest. In the early months of autumn, the Sun passes through the stars of Virgo, hence the connections with the gathering-in of fruit and wheat.

Spica is a hot, blue-white star over 12,000 times brighter than the Sun, boasting a temperature of 22,500°C. It has a stellar companion, which lies just 18 million kilometres away from Spica – closer than Mercury orbits the Sun. Both stars inflict a mighty gravitational toll on each other, raising enormous tides – creating two distorted, egg-shaped stars. In fact, Spica is the celestial equivalent of a rugby ball.

The glory of Virgo is the 'bowl' of the Y-shape. Scan it with a small telescope and you'll find it packed with faint, fuzzy blobs. These are just a few of the 2000 galaxies – star-cities like the Milky Way – that make up the gigantic **Virgo Cluster**.

PLANETS ON VIEW

Look low down in west after sunset and you'll see **Venus** starting to appear as the Evening Star: by the end of May, it's setting one-and-a-half hours after the Sun. At magnitude −3.8, Venus is bright enough to show up even against the evening twilight glow.

May, and 10 pm at the end of the month. The planets move slightly relative to the stars during the month.

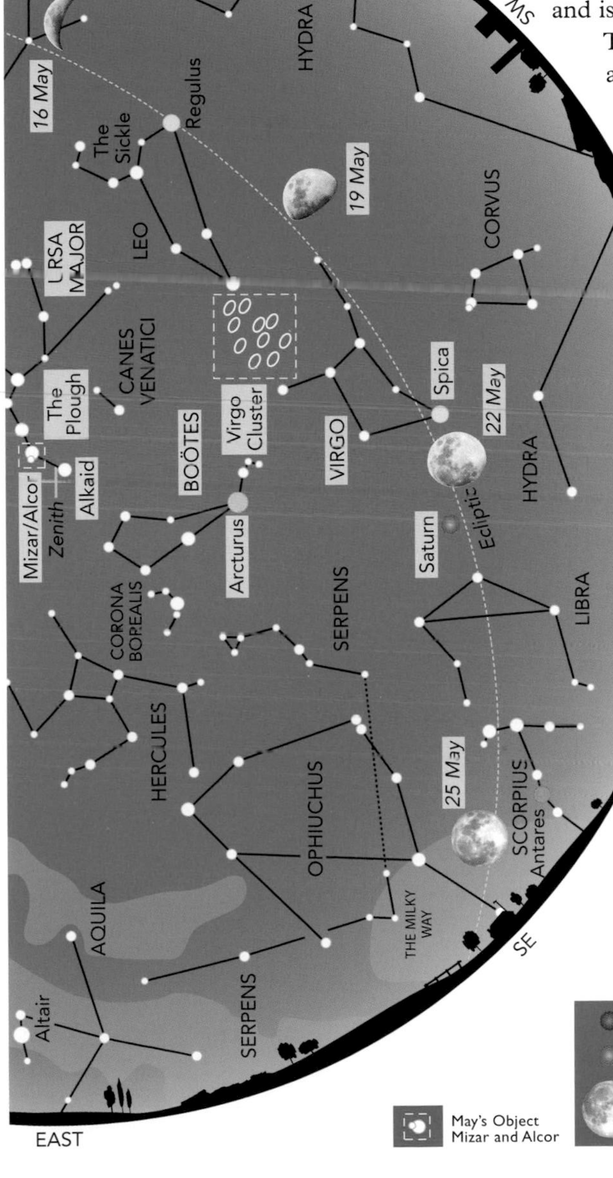

Jupiter
Saturn
Moon

Once Venus has set, and the sky is dark, **Jupiter** dominates the western sky. It's at magnitude −1.8, between the horns of the celestial bull, Taurus. Mid-month, the biggest planet sets around 11 pm.

During May, you'll see these two worlds approach each other, and pass only a degree apart on 28 May. It's not as spectacular as last year's conjunction of Venus and Jupiter, as both planets lie low down in the twilight. Look carefully with binoculars, and you can spot fainter **Mercury** (magnitude −0.5) just above the planetary pair.

On the other side of the sky, **Saturn** shines at magnitude +0.4 on the Libra/Virgo border, and is visible all night long.

Throughout May, **Mars**, **Uranus** and **Neptune** are lost in the Sun's glare.

MOON

The crescent Moon lies between Venus and Jupiter on 11 May, and it's next to Jupiter on 12 May (see Special Events). On 18 May, you'll find the First Quarter Moon close to Regulus. The Moon is near Spica on 21 May, and lies halfway between Spica and Saturn on 22 May.

SPECIAL EVENTS

The maximum of the Eta Aquarid meteor shower falls on **5/6 May**. It's an excellent year for viewing these tiny pieces of Halley's Comet burning up in Earth's atmosphere, as the Moon is well out of the way.

There's an annular solar eclipse on

MOON		
Date	Time	Phase
2	12.14 pm	Last Quarter
10	1.29 am	New Moon
18	5.35 am	First Quarter
25	5.25 am	Full Moon
31	7.58 pm	Last Quarter

10 May, visible from northern Australia and the south Pacific. The whole of Australasia and most of the Pacific will witness a partial eclipse, but nothing can be seen from the UK.

On **11 May**, look out for the slender crescent Moon low in the twilight around 9.30 pm, with Venus below and Jupiter above. The next day, **12 May**, the Moon forms a glorious pair with Jupiter.

Technically, the penumbral eclipse of the Moon (see April's Topic) at 5.10 am on **25 May** is visible from the UK, but it will be difficult to spot in the brightening dawn sky.

Viewing tip

It's best to view your favourite objects when they're well clear of the horizon. When you look low down, you're seeing through a large thickness of the atmosphere – which is always shifting and turbulent. It's like trying to observe the outside world from the bottom of a swimming pool! This turbulence makes the stars appear to twinkle. Low-down planets also twinkle – although to a lesser extent, because they subtend tiny discs, and aren't so affected.

MAY'S OBJECT

Home in on the 'kink' in the tail of **Ursa Major** (the Great Bear), and you'll spot the most famous pair of stars in the sky – **Mizar** (magnitude +2.4) and **Alcor** (magnitude +4.0). Generations of astronomers have referred to them as 'the horse and rider', and students have been raised on the fact that the pair make up a classic double star system, orbiting in each other's embrace. But *are* Mizar and Alcor an item? It seems not. Although they both lie about 80 light years away, they are separated by 3 light years – nearly the distance from the Sun to our closest star, Proxima Centauri. Undoubtedly, Mizar is a

complex star system, having a companion visible through a telescope which is itself double; in total, there are four stars involved. But it appears that Alcor is an innocent bystander. Although it shares its path through space with Mizar, the two are just members of the 'stellar association' of stars making up Ursa Major. Unlike most constellations, the stars of the Great Bear are genuinely linked by birth (with the exceptions of **Dubhe** and **Alkaid**, at opposite ends of the central '**Plough**'). So let's hear 'independence for Alcor'!

MAY'S PICTURE

As solar activity continues to hot up – as a result of our local star's 11-year magnetic cycle – look out for the Northern Lights. The **Aurora Borealis** is caused when energetic electrical particles from the Sun hit the Earth's magnetic poles. The result: a neon glow in the sky, and a glorious light-show of shifting curtains and rays. Although aurorae are normally restricted to high northern latitudes, very powerful displays can be seen further south.

◀ The Aurora Borealis (Northern Lights) seen over the Sunnmørs Alps mountains in Ørsta, Norway. This photo was taken by Geir T. Øye at about 1 am local time on 25 October 2011 using a Canon EOS 500D camera, 15 mm fisheye lens, tripod and remote controller. The exposure time was 23 seconds, ISO 1600.

MAY'S TOPIC
Saturn

The slowly moving ringworld Saturn is currently livening up the sprawling constellations of Virgo (the Virgin) and Libra (the Scales). It's famed for its huge engirdling appendages: the rings would stretch nearly all the way from the Earth to the Moon. The planet is a glorious sight through a small telescope, like an exquisite model hanging in space (see April's Picture).

And the rings are just the beginnings of Saturn's larger family. It has at least 62 moons, including Titan – which is also visible through a small telescope. The international Cassini–Huygens mission has discovered lakes of liquid methane and ethane on Titan, and possibly active volcanoes. And the latest exciting news is that Cassini has imaged plumes of salty water spewing from Saturn's icy moon Enceladus. Another moon, Dione, has traces of oxygen in its thin atmosphere. These discoveries raise the intriguing possibility of life on Saturn's moons.

Saturn itself is second only to Jupiter in size. But it's so low in density that were you to plop it in an ocean, it would float. Like Jupiter, Saturn has a ferocious spin rate – 10 hours and 32 minutes – and its winds roar at speeds of up to 1800 km/h.

Saturn's atmosphere is much blander than that of its larger cousin. But it's wracked with lightning-bolts 1000 times more powerful than those on Earth.

The Sun reaches its highest position over the northern hemisphere in June, so we get the longest days and shortest nights. The Summer Solstice this year takes place on 21 June, and the height of summer will be celebrated at festivals on this day – notably at Stonehenge, in Wiltshire.

This seasonal ritual traces its roots back through millennia, and has led to the construction of massive stone monuments aligned on the Sun at the solstices. Our ancestors clearly had formidable astronomical knowledge.

Take advantage of the soft, warm weather to acquaint yourself with the lovely summer constellations of **Hercules**, **Scorpius**, **Lyra**, **Cygnus** and **Aquila**.

▼ The sky at 11 pm in mid-June, with Moon positions at three-day intervals either side of Full Moon. The star positions are also correct for midnight at the beginning of

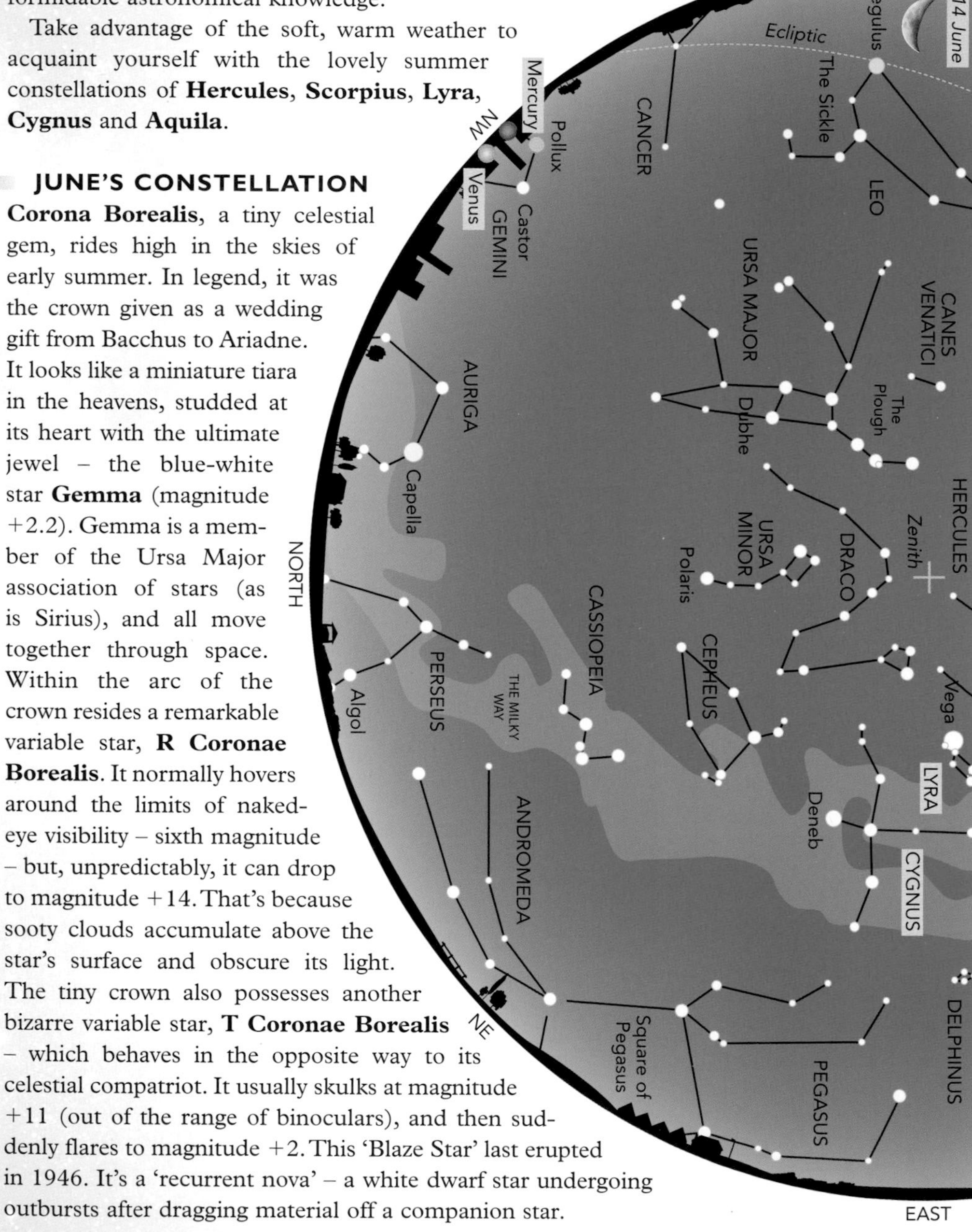

JUNE'S CONSTELLATION

Corona Borealis, a tiny celestial gem, rides high in the skies of early summer. In legend, it was the crown given as a wedding gift from Bacchus to Ariadne. It looks like a miniature tiara in the heavens, studded at its heart with the ultimate jewel – the blue-white star **Gemma** (magnitude +2.2). Gemma is a member of the Ursa Major association of stars (as is Sirius), and all move together through space. Within the arc of the crown resides a remarkable variable star, **R Coronae Borealis**. It normally hovers around the limits of naked-eye visibility – sixth magnitude – but, unpredictably, it can drop to magnitude +14. That's because sooty clouds accumulate above the star's surface and obscure its light. The tiny crown also possesses another bizarre variable star, **T Coronae Borealis** – which behaves in the opposite way to its celestial compatriot. It usually skulks at magnitude +11 (out of the range of binoculars), and then suddenly flares to magnitude +2. This 'Blaze Star' last erupted in 1946. It's a 'recurrent nova' – a white dwarf star undergoing outbursts after dragging material off a companion star.

June, and 10 pm at the end of the month. The planets move slightly relative to the stars during the month.

PLANETS ON VIEW

We have two 'evening stars' in the first half of June – but you'll need a low horizon to the north-west to see them. The brighter of these two planets, **Venus**, stays with us all month. At magnitude −3.8, the Planet of Love sets an hour-and-a-half after the Sun.

Look a few degrees to the upper left of Venus (binoculars will help a lot) and you'll see fainter **Mercury**, which is at its greatest eastern elongation on 12 June. You'll spot the innermost planet most easily at the start of June, when it's at magnitude −0.3; by 15 June it's faded away to magnitude +0.9.

Saturn (magnitude +0.6) lies in Virgo, setting around 2.30 am.

Mars, **Jupiter**, **Uranus** and **Neptune** are lost in the Sun's glare in June.

MOON

On 10 June, the thinnest of crescent Moons lies low down on the horizon to the lower left of Venus. The next evening – 11 June – it's to the left of Venus and Mercury (see Special Events). On 14 June, the Moon is below Regulus. The Moon passes less than a degree from Spica in the dusk on 18 June, and it lies near Saturn on 19 June.

SPECIAL EVENTS

On **11 June,** around 10.30 pm, look out for the crescent Moon on the north-western horizon with Venus well to its right – the 'star' between them is Mercury, while Castor and Pollux lie above.

21 June, 6.04 am: Summer Solstice. The Sun reaches its most northerly point in the sky, so 21 June is

June's Object Antares

June's Picture Lagoon and Trifid Nebulae

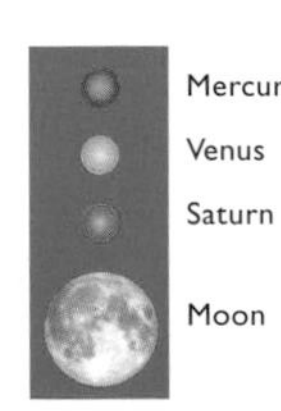

MOON		
Date	Time	Phase
8	4.56 pm	New Moon
16	6.24 pm	First Quarter
23	12.32 pm	Full Moon
30	5.53 am	Last Quarter

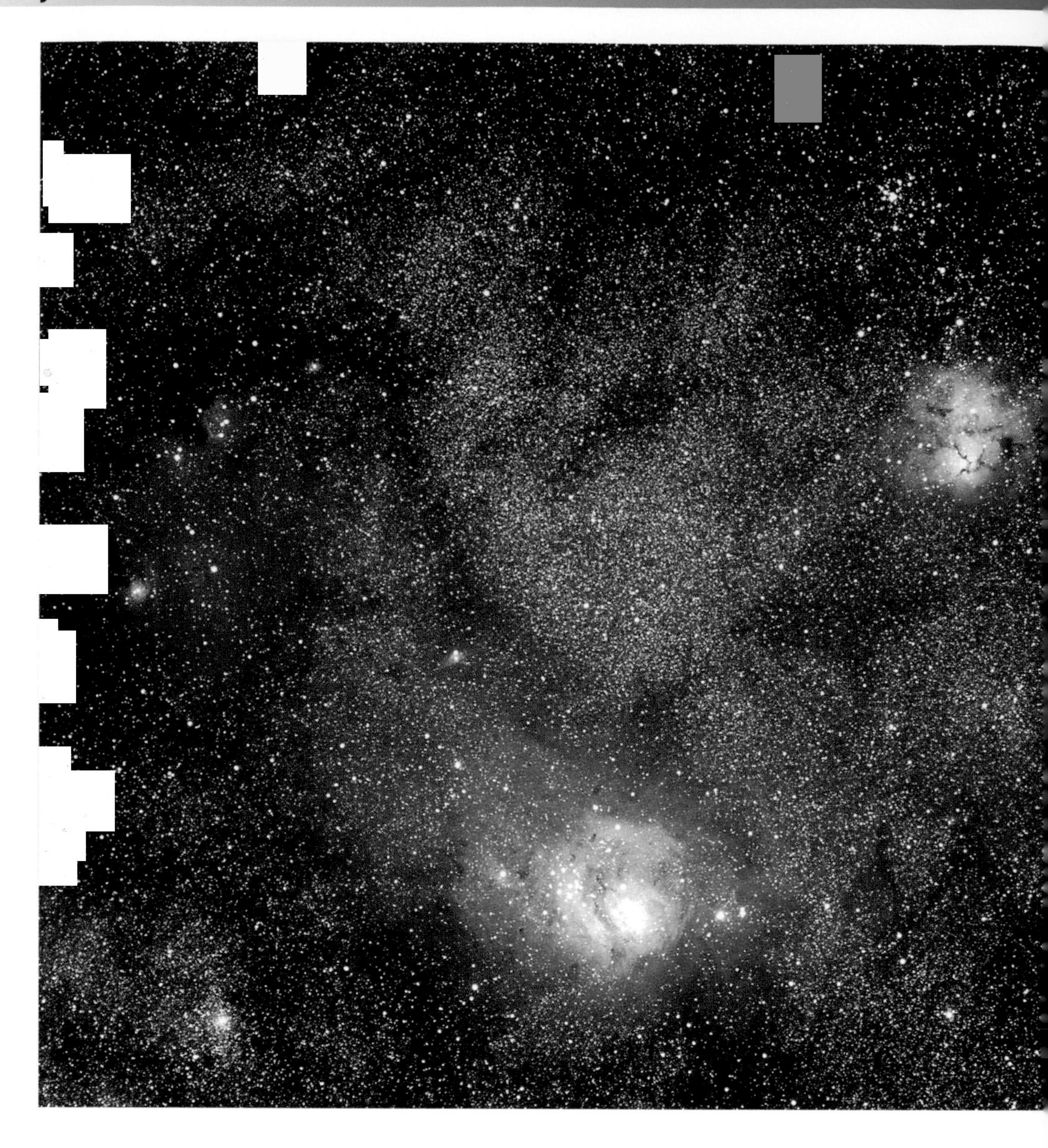

Midsummer's Day, with the longest period of daylight. Correspondingly, we have the shortest nights.

We are treated to an unusually large and romantic 'Moon in June' on **23 June**. That's partly because June's Full Moon is always near the horizon, which makes it look bigger. But on that date, the Moon also happens to be at its closest to the Earth this year – 357,000 kilometres away – so it's almost 30% bigger and brighter than when the Moon is at its farthest point.

▲ Michael Stecker's shot was taken through an Astrophysics 130 mm f/6 refractor with a Pentax 6 × 7 film camera. This is a composite of two 45-minute exposures on Kodak Pro 400 PPF film.

JUNE'S OBJECT

Its name means 'the rival of Mars' – and you can see why! Look low in the south this month for a baleful red star that marks the heart of the constellation of **Scorpius**. Six hundred light years away, **Antares** is a bloated red giant star near the end of its life. Running out of its central supplies of nuclear fuel, its core has shrunk and heated up, causing its outer layers to billow out and cool. Antares is at least 15 times heavier than our Sun, and now it's grown to over 500 times bigger and 10,000 times more luminous. Placed at the centre of our Solar System, it would engulf all the planets out to Mars. And its size isn't constant: Antares' gravity hasn't got to grips with its extended girth, making the star swell and shrink, changing in brightness as it does so. The giant star has a small blue companion (magnitude +5), which is hard to see against Antares' glare. Just visible in a small telescope, the star circles the red giant every 878 years. Eventually, the core of Antares will collapse completely, and it will explode as a brilliant supernova.

Viewing tip

This is the month for the best Sun-viewing – but be careful. **NEVER** use a telescope or binoculars to look at the Sun directly: it could blind you permanently. Fogged film is no safer, because it allows the Sun's infra-red (heat) rays to get through. Eclipse goggles are safe (unless they're scratched). The best way to observe the Sun is to project its image through binoculars or a telescope on to a white piece of card. Now is the time to catch up with our local star: its sunspot activity (and associated flares and eruptions) is approaching its (roughly) 11-year maximum.

JUNE'S PICTURE

The constellation of **Sagittarius** is a hot-spot for nebulae. When you look at the constellation, you're gazing towards the centre of our Milky Way, and the young gas and dust clouds seem to pile up upon one another. To the right in this image is the compact **Trifid Nebula** (M20), dissected by drifts of dust; below is the larger **Lagoon Nebula** (M8). It measures over 100 light years in diameter, and is one of only two starbirth regions visible to the unaided eye from the northern hemisphere (the other being the Orion Nebula).

JUNE'S TOPIC
Noctilucent clouds

Look north at twilight, and you may be lucky enough to see what has to be the most ghostly apparition in the night sky – noctilucent clouds. Derived from the Latin 'night shining', these spooky clouds glow blue-white. Illuminated by the Sun from below the horizon, they're most commonly seen between latitudes 50° and 70° in the summer, when the Sun towers over the northern hemisphere.

These are the highest clouds in the sky, occurring around 80 kilometres up in the atmosphere. And their origin is controversial. They're certainly composed of ice, coated around tiny particles of dust – but what is the nature of the dust?

Tellingly, the first observation of noctilucent clouds was made in 1885, two years after the eruption of Krakatoa. So could the particles be volcanic dust? Others believe that the dust could be micrometeorites, entering the atmosphere at high altitudes. Some scientists put them down to the industrial revolution, with its resultant increased pollution.

Forget planets – this July is a month for stars! Gaze at the **Summer Triangle** as it soars overhead. **Vega**, **Deneb** and **Altair** are each the brightest star in their own constellation: Vega in **Lyra**, Deneb in **Cygnus** and Altair in **Aquila**. And this is the time to catch the far-southern constellations of **Sagittarius** and **Scorpius** – embedded in the glorious heart of the Milky Way.

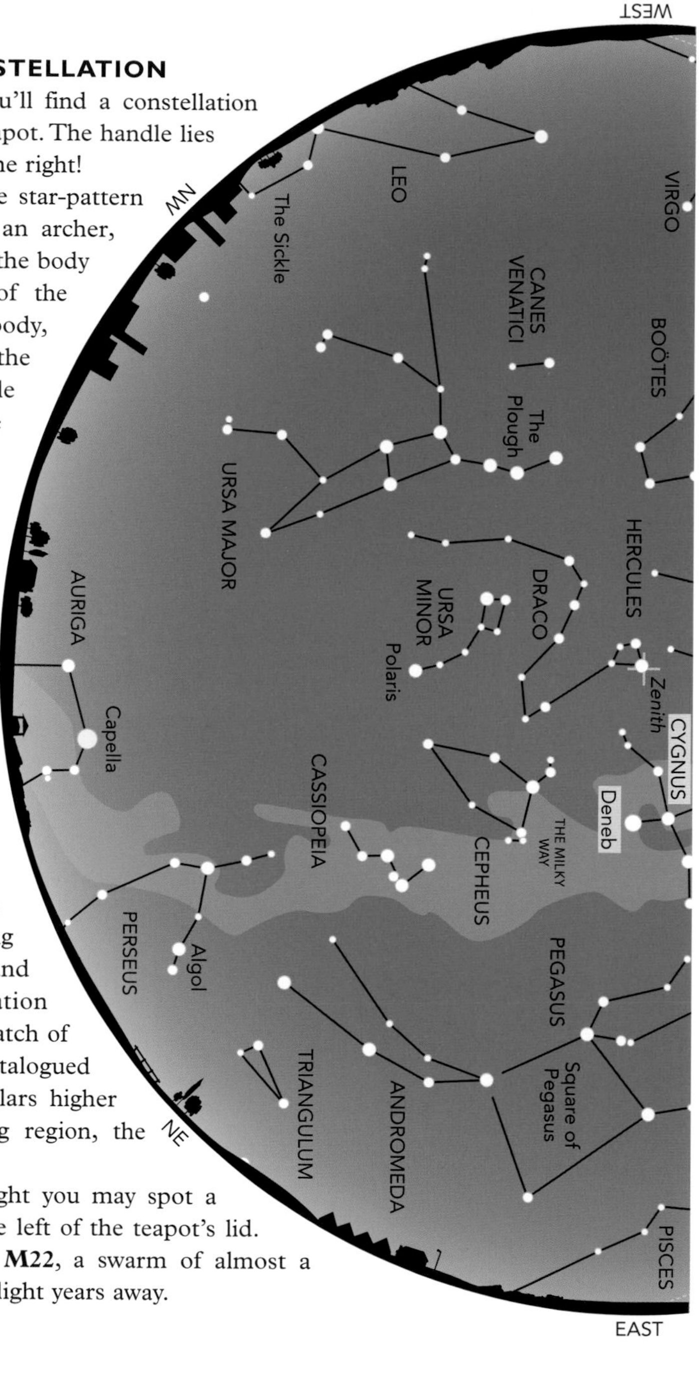

▼ The sky at 11 pm in mid-July, with Moon positions at three-day intervals either side of Full Moon. The star positions are also correct for midnight at the beginning of

JULY'S CONSTELLATION

Low down in the south, you'll find a constellation that's shaped rather like a teapot. The handle lies to the left and the spout to the right!

To the ancient Greeks, the star-pattern of **Sagittarius** represented an archer, with the torso of a man and the body of a horse. The 'handle' of the teapot represents his upper body, the curve of three stars to the right are his bent bow, while the end of the spout is the point of the arrow, aimed at Scorpius, the fearsome celestial scorpion.

Sagittarius is rich in nebulae and star clusters. If you have a clear night (and preferably from a southern latitude), sweep Sagittarius with binoculars for some fantastic sights. Above the spout lies the wonderful **Lagoon Nebula** (see June's Picture) – visible to the naked eye on clear nights. This is a region where stars are being born. Between the teapot and the neighbouring constellation Aquila, you'll find a bright patch of stars in the Milky Way (catalogued as **M24**). Raise your binoculars higher to spot another star-forming region, the **Omega Nebula**.

Finally, on a very dark night you may spot a fuzzy patch, above and to the left of the teapot's lid. This is the globular cluster **M22**, a swarm of almost a million stars that lies 10,000 light years away.

July, and 10 pm at the end of the month. The planets move slightly relative to the stars during the month.

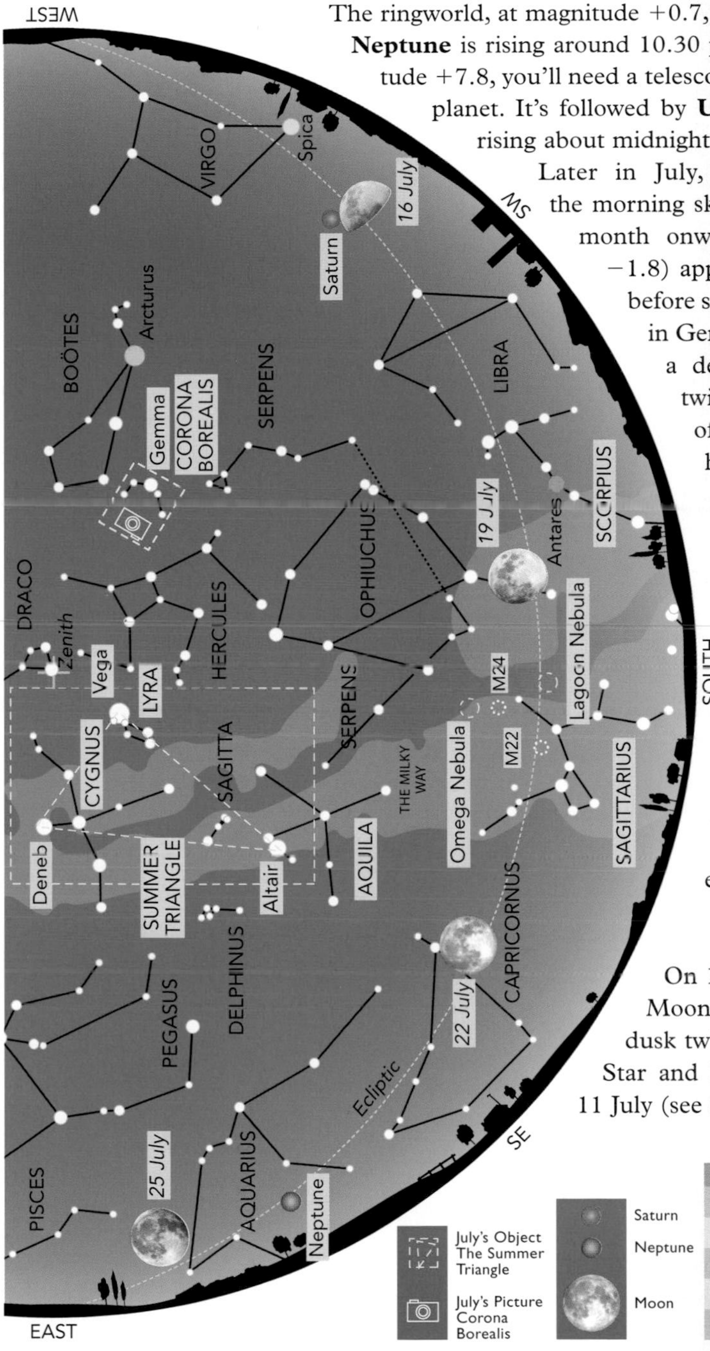

PLANETS ON VIEW

Venus (magnitude −3.8) skulks in the evening twilight all month in the north-north-west, setting one-and-a-half hours after the Sun.

After sunset, you'll find **Saturn** in the south-west in Virgo, less than a degree from the star kappa Virginis (magnitude +4.1). The ringworld, at magnitude +0.7, is setting about 0.30 am.

Neptune is rising around 10.30 pm in Aquarius. At magnitude +7.8, you'll need a telescope to spot the most distant planet. It's followed by **Uranus** (magnitude +5.8), rising about midnight in Pisces.

Later in July, the action picks up in the morning sky. From the middle of the month onwards, **Jupiter** (magnitude −1.8) appears low in the north-east before sunrise. The giant planet lies in Gemini, and it passes less than a degree below **Mars** in the twilight glow on the morning of 22 July; binoculars will help in spotting the Red Planet, as it's 20 times fainter than Jupiter, at magnitude +1.6.

From 27 July until the end of the month, **Mercury** appears very low in the north-east, 10 degrees to the lower left of Jupiter, and brightening from magnitude +0.8 to +0.2. The innermost planet is at greatest western elongation on 30 July.

MOON

On 10 July the slender crescent Moon lies below Venus in the dusk twilight; it's near the Evening Star and Regulus the next evening, 11 July (see Special Events). The Moon

WEST

EAST

SOUTH

July's Object The Summer Triangle

July's Picture Corona Borealis

Saturn

Neptune

Moon

MOON		
Date	Time	Phase
8	8.14 am	New Moon
16	4.18 am	First Quarter
22	7.15 pm	Full Moon
29	6.43 pm	Last Quarter

◀ *This photograph of Corona Borealis was taken by Robin Scagell on Ektachrome 1600 film using a 55 mm lens and exposure time of 5 minutes. A diffusion filter has been used to emphasize the star colours.*

is near Spica on 15 July. On 16 July, the Moon lies below Saturn. The star near the Moon on 18 and 19 July is Antares.

SPECIAL EVENTS

On **5 July**, at 7.59 pm, the Earth reaches aphelion, its furthest point from the Sun – 152 million kilometres out.

Look low in the north-western twilight around 10 pm on **11 July** for the lovely sight of the crescent Moon and Venus, with Regulus lying just above.

JULY'S OBJECT

The **Summer Triangle** is very much part of this season's skies (and it hangs around for most of the autumn, too!). It's made up of Vega, Deneb and Altair – the brightest stars in the constellations of Lyra (the Lyre), Cygnus (the Swan) and Aquila (the Eagle), respectively. The trio of stars make a striking pattern almost overhead on July nights.

The stars may seem to be almost the same brightness, but they're very different beasts. **Altair** – its name means 'flying eagle' – is one of the Sun's nearest neighbours, at a distance of nearly 17 light years. It's about ten times brighter than the Sun and spins at a breakneck rate of once every nine hours – as compared to around 30 days for our local star.

Vega, just over 25 light years away, is a brilliant white star

Viewing tip

This is the month when you really need a good, unobstructed view to the southern horizon to make out the summer constellations of Scorpius and Sagittarius. They never rise high in temperate latitudes, so make the best of a southerly view – especially over the sea – if you're away on holiday. A good southern horizon is also best for views of the planets, because they rise highest when they're in the south.

nearly twice as hot as the Sun. In 1850, it was the first star to be photographed. Now, more sensitive instruments have revealed that Vega is surrounded by a dusty disc – which may be a planetary system in the process of formation.

While **Deneb** – meaning 'tail' (of the swan) – may appear to be the faintest of the trio, the reality is different. It lies about 1500 light years away – because the star is so distant, these measurements will always be controversial, so watch this space! To appear so bright in our skies, it must be truly luminous. It's estimated that Deneb is over 60,000 times brighter than our Sun, and is one of the most brilliant stars known.

JULY'S PICTURE

The tiara-shaped constellation **Corona Borealis** (the Northern Crown – see June's Constellation) is one of two celestial coronets – its twin, Corona Australis (the Southern Crown), lies below Sagittarius and isn't visible from Britain. Despite its striking shape, this constellation isn't a real arc of stars: it's a chance arrangement of stars at very different distances. The jewel in the crown, **Gemma**, lies only 78 light years from us, while the other stars range from 114 to 350 light years away

JULY'S TOPIC
The Sun

The Sun is pretty lively now – and in more ways than one. At the height of summer, our local star rides high in the sky, and we feel the heat of its rays. Some 150 million kilometres away, the Sun is our local star – and our local nuclear reactor.

This giant ball of hydrogen gas is a vast hydrogen bomb. At its core, where temperatures reach 15.7 million degrees, the Sun fuses atoms of hydrogen into helium. Every second, it devours 4 million tonnes of itself, bathing the Solar System with light and warmth.

But the Sun is also a dangerous place. It has emerged from a quiet period to become active again, a cycle that repeats roughly every 11 years. The driver is the Sun's magnetic field, wound up by the spinning of our star's surface gases. The magnetic activity suppresses the Sun's circulation, leading to a rash of dark sunspots. Then the pent-up energy is released in a frenzy of activity, when our star hurls charged particles through the Solar System. These dangerous particles can kill satellites, and disrupt power lines on Earth. And if we are ever to make the three-year human journey to Mars, we will have to take the Sun's unpredictable, malevolent weather into account.

We have the Glorious Twelfth for astronomers this month: 12 to 13 August is the maximum of the **Perseid** meteor shower. This is the year's most reliable display of shooting stars, and it also conveniently takes place during the summer, when it's not too uncomfortable to stay up late under the stars!

▼ The sky at 11 pm in mid-August, with Moon positions at three-day intervals either side of Full Moon. The star positions are also correct for midnight

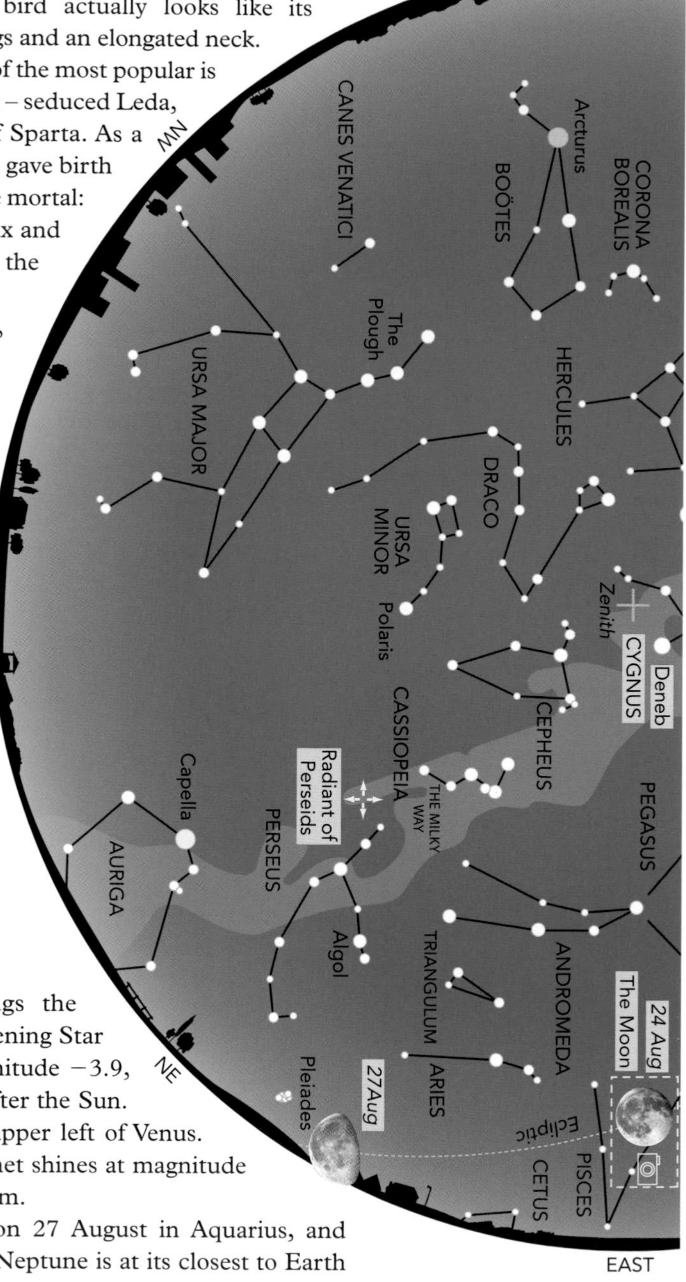

AUGUST'S CONSTELLATION

The flying swan, **Cygnus**, is one of our most-cherished summer constellations. The celestial bird actually looks like its namesake, with outspread wings and an elongated neck.

Swan legends abound. One of the most popular is that Zeus – disguised as a swan – seduced Leda, the wife of King Tyndareus of Sparta. As a result, the unfortunate woman gave birth to twins, one immortal and one mortal: they appear in the sky as Pollux and Castor, the heavenly twins in the constellation of Gemini.

Deneb (see July's Object), forms the swan's tail. The head is marked by what's probably the most beautiful double star in the sky: **Albireo**. You can spot the pair with good binoculars, but the combination of the pair of blue and gold stars is sensational when seen through a small telescope.

The Milky Way meanders through Cygnus. Edge-on, the band of our Galaxy is riddled with star clusters and nebulae. It's glorious to sweep Cygnus with binoculars or a small telescope.

PLANETS ON VIEW

All month long, **Venus** hugs the western horizon as a pallid Evening Star in the bright twilight. At magnitude −3.9, it's setting just over an hour after the Sun.

You'll find **Saturn** to the upper left of Venus. Lying in Virgo, the ringed planet shines at magnitude +0.8 and sets around 10.30 pm.

Neptune is at opposition on 27 August in Aquarius, and visible all night long. Though Neptune is at its closest to Earth

at the beginning of August, and 10 pm at the end of the month. The planets move slightly relative to the stars during the month.

this month, you need telescope power to track down this most distant planet, at a dim magnitude +7.8.

Uranus (magnitude +5.8) lies in Pisces and rises about 10 pm.

Hang on until 2 am, and you'll see brilliant **Jupiter** rising in the east. The giant planet, in Gemini, shines at magnitude −1.8.

At the beginning of the month, **Mars** lies just a few degrees to the lower left of Jupiter – rising at the same time – but it gradually pulls away during August. The Red Planet (magnitude +1.6) passes below the twin stars of Gemini, Castor and Pollux, in the last week of the month and moves into Cancer

For the first half of August, **Mercury** is a lovely morning star, rising two hours before the Sun in the north-east, and brightening from magnitude 0.0 to −1.2.

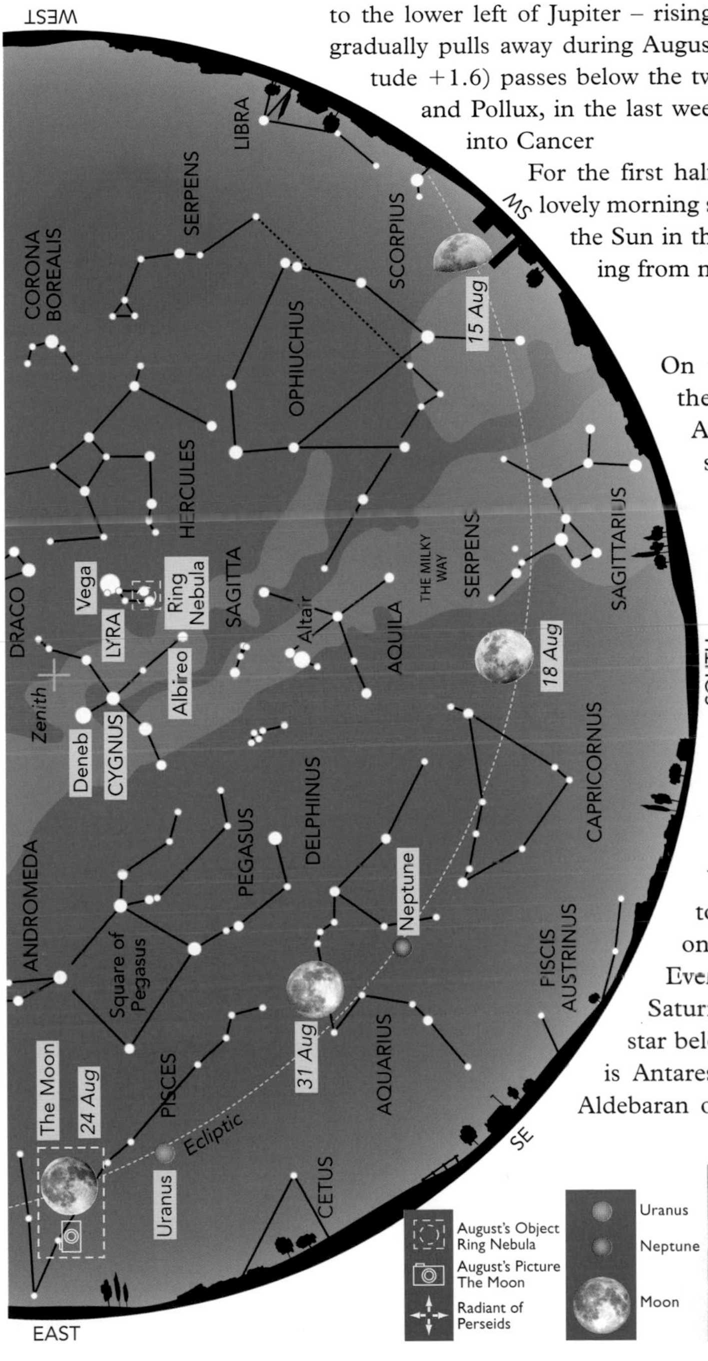

MOON

On the morning of 1 August, the Moon passes above Aldebaran. In the morning skies of 3 August, the crescent Moon lies to the right of Jupiter. At dawn the next day, 4 August, the Moon has a rendezvous with Jupiter, Mars and Mercury (see Special Events). And just before sunrise on 5 August, the slenderest of crescent Moons lies to the right of Mercury. The waxing crescent Moon is beneath Venus on 9 August, and lies to the left of the Evening Star on 10 August (see Special Events). The Moon lies near Saturn on 12 and 13 August. The star below the Moon on 15 August is Antares. The Moon is back with Aldebaran on the mornings of 28 and

MOON		
Date	Time	Phase
6	10.51 pm	New Moon
14	11.56 am	First Quarter
21	2.45 am	Full Moon
28	10.35 am	Last Quarter

29 August, while the morning of 31 August sees the Moon to the right of Jupiter.

SPECIAL EVENTS

On **4 August**, around 4 am, you'll find the crescent Moon making a striking triangle with Jupiter and Mars, with Mercury well to the lower left.

The waxing crescent Moon appears low in the evening sky with the brilliant Evening Star, Venus, on **10 August**.

The maximum of the annual **Perseid** meteor shower falls on **12/13 August**. Look forward to spectacular views after midnight, when the Moon has set and the sky is really dark. Expect to see about one meteor per minute.

AUGUST'S OBJECT

Tucked into the small constellation of **Lyra** (the Lyre) – near the brilliant star **Vega** – lies a strange celestial sight. It was first spotted by French astronomer Antoine Darquier in 1779, as 'a very dull nebula, but perfectly outlined; as large as Jupiter and looks like a fading planet.' Under higher magnification, it appears as a bright ring of light with a dimmer centre. Hence its usual name, the **Ring Nebula**.

Viewing tip
Have a Perseids party! You don't need any optical equipment – in fact, telescopes and binoculars will restrict your view of the meteor shower. The ideal viewing equipment is your unaided eye, plus a sleeping bag and a lounger on the lawn. If you want to make measurements, a stopwatch and clock are good for timings, while a piece of string will help to measure the length of the meteor trail.

You can make out the Ring Nebula with even a small telescope, though it's so compact that you'll need a magnification of over 50 times to distinguish it from a star.

The Ring Nebula is the remains of a dying star – a cloud of gas lit up by the original star's incandescent core. For centuries, astronomers assumed that the Ring Nebula was a sphere of gas. But the Hubble Space Telescope has now found that it's actually barrel-shaped. It looks like a ring to us only because we happen to view the barrel end-on. Aliens observing the Ring Nebula from another perspective would undoubtedly call it something different!

AUGUST'S PICTURE

Our near-neighbour, the **Moon** – pockmarked and battle-scarred by years of bombardment. The dark 'seas' (which make up the 'face' of the 'Man in the Moon') were created by asteroid impacts some 3.8 billion years ago, when lava from the lunar interior welled up to fill the scars. This image reveals the subtle colour variations in the 'seas', due to differences in the minerals making up the lava. The crater at the bottom is Tycho, 85 kilometres across. Its fresh appearance and brilliant rock ejecta lead astronomers to believe that Tycho is one of the Moon's youngest features.

◀ *Gibbous Moon, photographed with a 200 mm Sky-Watcher reflector and Canon 550D camera by James Dyson from Warrington in Cheshire.*

AUGUST'S TOPIC
Constellations

Cygnus, our constellation of the month, highlights our obsession to 'join up the dots' in the sky and weave stories around them. But why have we done this? One explanation is that – because the constellations on view changed during the year, as the Earth moved around the Sun – the patterns acted as an aide memoire to where we were in our annual cycle. This would have been especially useful to ancient farming communities.

Another reason is that the stars were a great steer to navigation at sea. Scholars believe that Greek astronomers 'mapped' their legends on to the sky, so that sailors crossing the Mediterranean would associate certain constellations with their traditional stories.

However, not all the world saw the heavens through western eyes. The Chinese divided up the sky into a plethora of tiny constellations, containing three or four stars apiece. And the Australian Aborigines, in their dark deserts, were so overwhelmed with stars that they made constellations out of dark patches where they couldn't see any!

Autumn is here – with its unsettled weather – and we have wet star patterns to match! **Aquarius** (the Water Carrier) is part of a group of aqueous star patterns that includes **Cetus** (the Sea Monster), **Capricornus** (the Sea Goat), **Pisces** (the Fishes), **Piscis Austrinus** (the Southern Fish) and **Delphinus** (the Dolphin). There's speculation that the ancient Babylonians associated this region with water because the Sun passed through this zone of the heavens during their rainy season, from February to March.

▼ The sky at 11 pm in mid-September, with Moon positions at three-day intervals either side of Full Moon. The star positions are also correct for midnight at

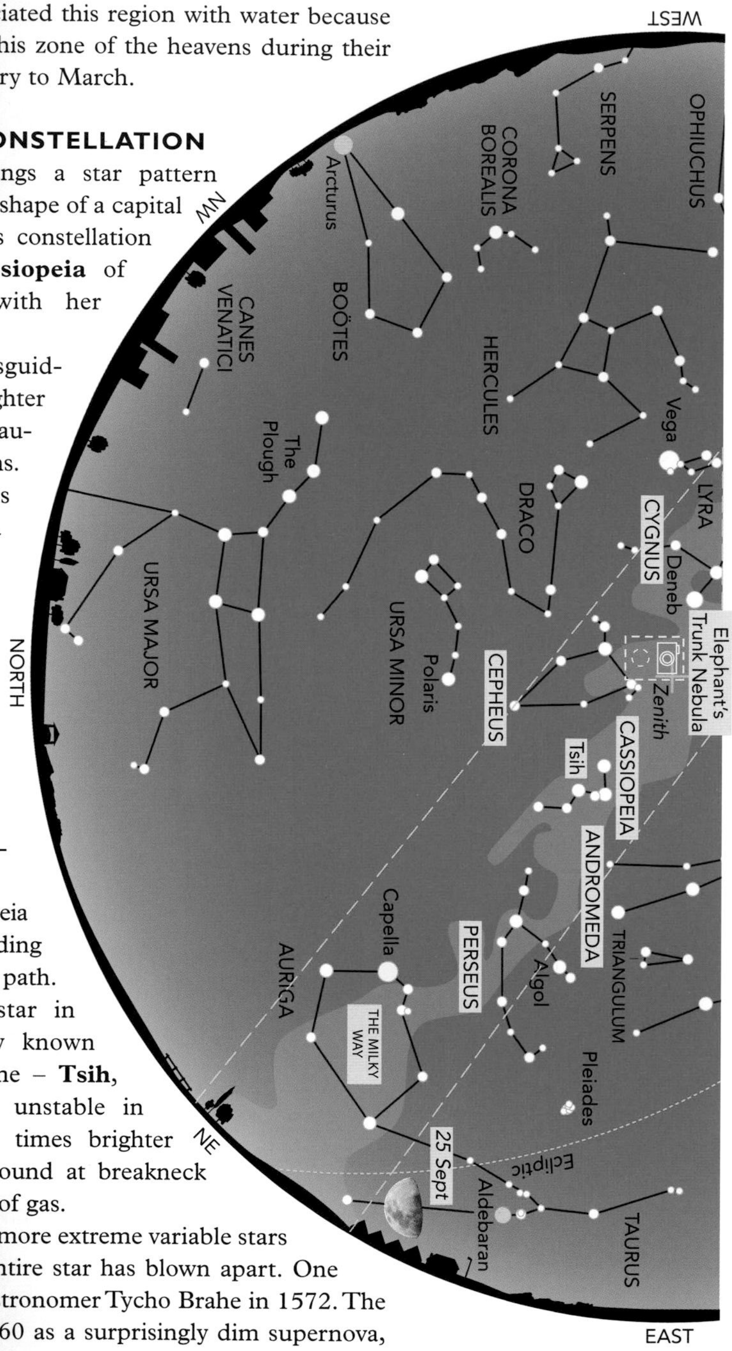

SEPTEMBER'S CONSTELLATION

In the northern sky hangs a star pattern forming the unmistakable shape of a capital 'W'. To the ancients, this constellation represented Queen **Cassiopeia** of Ethiopia, who ruled with her husband King **Cepheus**.

Queen Cassiopeia misguidedly boasted that her daughter **Andromeda** was more beautiful than the sea nymphs. Poseidon, the sea god, was so incensed that he sent a ravaging monster (**Cetus**) to eat the young people of the country. It could only be appeased by the sacrifice of Andromeda – but she was rescued by the hero **Perseus**. Cassiopeia, Cepheus, Andromeda, Perseus and Cetus are now all immortalized in the heavens.

The Chinese saw Cassiopeia as three star groups, including a chariot and a mountain path. Unusually, the central star in Cassiopeia is universally known today by its Chinese name – **Tsih**, 'the whip'. This star is unstable in brightness. Some 55,000 times brighter than the Sun, it spins around at breakneck pace, flinging out streams of gas.

Cassiopeia has seen two more extreme variable stars – supernovae, where an entire star has blown apart. One was observed by Danish astronomer Tycho Brahe in 1572. The other exploded around 1660 as a surprisingly dim supernova,

the beginning of September, and 10 pm at the end of the month. The planets move slightly relative to the stars during the month.

but its expanding gases form the most prominent radio source in the sky, Cassiopeia A.

PLANETS ON VIEW

We have an Evening Star in September. But **Venus** is very hard to spot as it's setting little more than an hour after the Sun: it's almost lost in the western twilight glow, even though Venus shines brighter than any star at magnitude −4.0.

Saturn is visible higher up in the evening sky, on the border of Virgo and Libra. Shining at magnitude +0.8, the ringed planet sets just before 10 pm at the start of September. But it's slipping downwards, and – after passing four degrees above Venus on 17 September – Saturn disappears into the twilight.

At magnitude +7.8, faint and distant **Neptune** lies in Aquarius, setting around 5 am. You'll find **Uranus** (magnitude +5.7) in Pisces, rising just before 8 pm.

Just after midnight, magnificent **Jupiter** rises in the north-east, in Gemini, shining at magnitude −2.0. It's followed by **Mars** (magnitude +1.6), which rises about 2.30 am and late in the month moves from Cancer into Leo. On the morning of 9 September, the Red Planet passes right in front of Praesepe (see Special Events).

Mercury is too close to the Sun to be visible this month.

MOON

The crescent Moon lies below Jupiter in the morning of 1 September, with

September's Object
The Milky Way

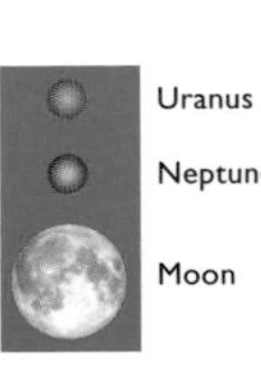

MOON		
Date	Time	Phase
5	12.36 pm	New Moon
12	6.08 pm	First Quarter
19	12.13 pm	Full Moon
27	4.56 am	Last Quarter

Mars to the lower left. The next morning, 2 September, the Moon is next to the Red Planet, and Jupiter is high above. Back in the evening twilight sky, the crescent Moon lies immediately below Venus on 8 September (see Special Events). You'll find the Moon just to the lower left of Saturn on 9 September. On 11 September, the Moon lies above Antares. The night of 24/25 September sees the Moon passing through the fringes of the Hyades, next to Aldebaran. On the morning of 28 September, the Moon is to the right of Jupiter; the following morning, 29 September, it has moved below the giant planet.

► *Part of the nebula IC1396, known as the Elephant's Trunk, photographed with a 106 mm refracting telescope and CCD camera by Gordon Rogers from Long Crendon in Buckinghamshire. Separate exposures were made through narrowband filters to produce this false-colour image.*

SPECIAL EVENTS

On **8 September**, the crescent Moon makes a striking sight with Venus, very low in the west around 8 pm.

In the morning of **9 September**, Mars moves right in front of Praesepe, the star cluster commonly known as the Beehive – a gleaming ruby in a jewel-box of stars, and a lovely sight in binoculars or a small telescope.

It's the Autumn Equinox at 9.44 pm on **22 September**. The Sun is over the Equator as it heads southwards in the sky, and day and night are equal.

SEPTEMBER'S OBJECT

It's a stunning month for sweeping down the **Milky Way**, especially through binoculars. The stars look packed together, and you'll pick out star clusters and nebulae as you travel its length. These are all the more distant denizens of our local Galaxy, flattened into a plane because we live within its disc. It's akin to seeing the overlapping streetlights of a distant city on Earth.

But you'll notice something else – there is a black gash between the stars. William Herschel, the first astronomer to map the Galaxy, thought that this was a hole in space. But now we know that the **Great Rift** (in **Cygnus** – see August's Constellation) is a dark swathe of sooty dust crossing the disc of our Galaxy. It is material poised to collapse under gravity, heat up, and – mixed with interstellar gas – create new generations of stars and planets. Proof that there is life in our old Galaxy yet!

Viewing tip

Around 2.5 million light years away from us, the Andromeda Galaxy is often described as the furthest object easily visible to the unaided eye. But it's not that easy to see – especially if you are suffering from light pollution. The trick is to memorize the star patterns in Andromeda and look slightly to the side of where you expect the galaxy to be. This technique – called 'averted vision' – causes the image to fall on parts of the retina that are more light-sensitive than the central region, which is designed to see fine detail.

SEPTEMBER'S PICTURE

The **Elephant's Trunk Nebula** in **Cepheus** gets its name from a curving dark cloud of dust and gas illuminated by young stars being born inside. Some 2400 light years away, the nebula has given birth to several stars less than 100,000 years old (bear in mind that the Sun is 4.6 *billion* years old). The nebula is a hotbed of activity: violent winds from the young stars stir up the gas and fuel new starbirth.

SEPTEMBER'S TOPIC
The Harvest Moon

It's the time of year when farmers work late into the night, bringing home their ripe crops before autumn sets in. And traditionally they are aided by the light of the 'Harvest Moon' – a huge glowing Full Moon that seems to hang constant in the evening sky, rising at almost the same time night after night. At first sight, that doesn't seem possible. After all, the Moon is moving around the Earth, once in just under a month, so it ought to rise roughly an hour later every night. But things in the sky are hardly ever that simple....

The Moon follows a tilted path around the sky (close to the line of the ecliptic, which is marked on the star chart). And this path changes its angle with the horizon at different times of year. On September evenings, the ecliptic runs roughly parallel to the horizon, so night after night the Moon moves to the left in the sky, but it hardly moves downwards. As a consequence, the Moon rises around the same time for several consecutive nights. This year, the Full Moon on 19 September rises at 6.48 pm (ideal for harvesting); it rises just 25 minutes earlier the evening before, and 26 minutes later the night after.

The glories of October's skies can best be described as 'subtle'. The barren square of **Pegasus** dominates the southern sky, with **Andromeda** attached to his side. But there are two glorious galaxies on view: the **Andromeda Galaxy** (see this month's Picture) and its smaller cousin **M33** in **Triangulum** (see this month's Object). When you gaze upon these cosmic beauties, you are looking back over 2 million years in time.

▼ *The sky at 11 pm in mid-October, with Moon positions at three-day intervals either side of Full Moon. The star positions are also correct for midnight at*

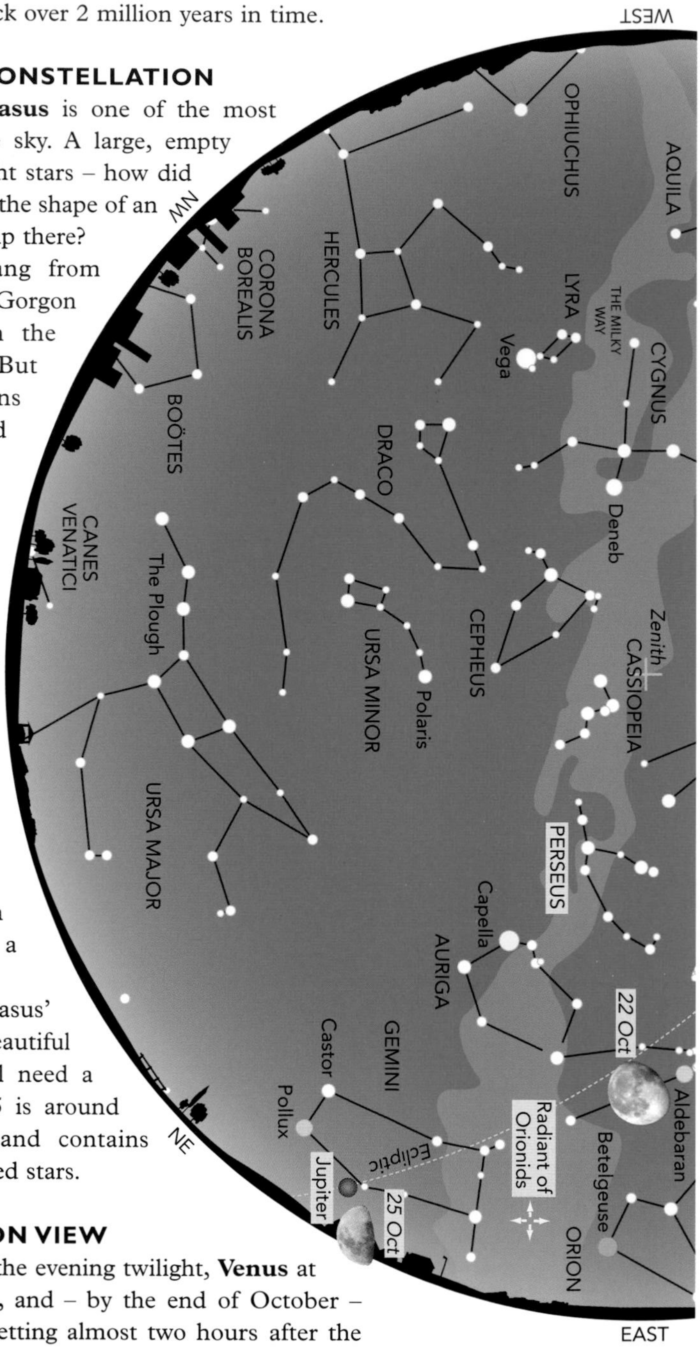

WEST
OPHIUCHUS
AQUILA
NW
CORONA BOREALIS
HERCULES
LYRA
THE MILKY WAY
Vega
CYGNUS
BOÖTES
DRACO
Deneb
CANES VENATICI
The Plough
CEPHEUS
Zenith
CASSIOPEIA
NORTH
URSA MINOR
Polaris
URSA MAJOR
PERSEUS
Capella
AURIGA
22 Oct
Castor
GEMINI
Pollux
Aldebaran
Radiant of Orionids
Betelgeuse
NE
Ecliptic
Jupiter
25 Oct
ORION
EAST

OCTOBER'S CONSTELLATION

It has to be said that **Pegasus** is one of the most boring constellations in the sky. A large, empty square of four medium-bright stars – how did our ancestors manage to see the shape of an upside-down winged horse up there?

In legend, Pegasus sprang from the blood of Medusa the Gorgon when **Perseus** (nearby in the sky) severed her head. But all pre-classical civilizations have their fabled winged horse, and we see them depicted on Etruscan and Euphratean vases.

The star at the top right of the square – **Scheat** – is a red giant over a hundred times wider than the Sun. Close to the end of its life, it pulsates irregularly, changing in brightness by about a magnitude. **Enif** ('the nose') – outside the square to the right – is a yellow supergiant. A small telescope, or even good binoculars, will reveal a faint blue companion star.

Just next to Enif – and Pegasus' best-kept secret – is the beautiful globular cluster **M15**. You'll need a telescope for this one. M15 is around 30,000 light years away, and contains about 100,000 densely-packed stars.

PLANETS ON VIEW

After months of loitering in the evening twilight, **Venus** at last pushes its way upwards, and – by the end of October – you'll see it in a dark sky, setting almost two hours after the

the beginning of October, and 9 pm at the end of the month (after the end of BST). The planets move slightly relative to the stars during the month.

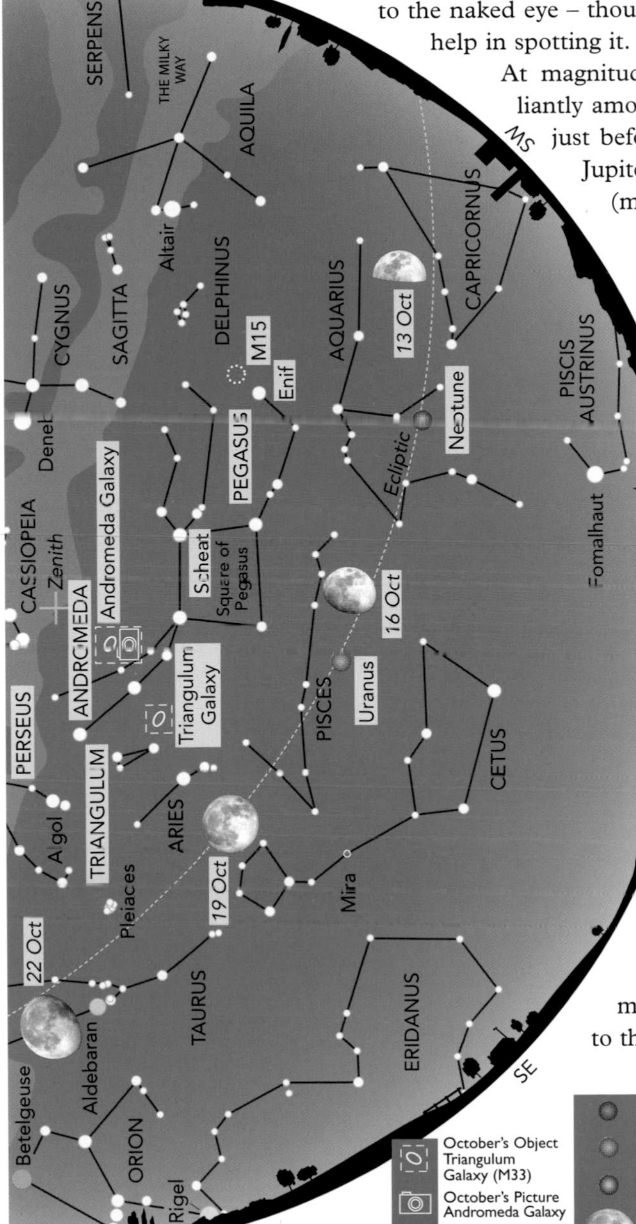

Sun. The Evening Star brightens from magnitude −4.1 to −4.3 during the month. Through a telescope, you'll see the planet's shape change from gibbous (three-quarter lit) to half-lit as Venus swings round the Sun.

Neptune (magnitude +7.9) lies in Aquarius, and sets about 3 am. Its brighter sibling, **Uranus**, is at opposition on 3 October, so this is as good a view as we'll ever get! The seventh planet lies in Pisces, and at magnitude +5.7 it's just visible to the naked eye – though binoculars are certainly a great help in spotting it.

At magnitude −2.1, **Jupiter** is shining brilliantly among the stars of Gemini, and rising just before 11 pm. The star very close to Jupiter at the start of October is Wasat (magnitude +3.6).

Mars (magnitude +1.6) rises around 2.30 am. During the month, the Red Planet steams through Leo, passing just a degree above Regulus on the morning of 15 October.

Saturn and **Mercury** are lost in the Sun's glare in October, even though the innermost planet is at greatest eastern elongation on 9 October.

MOON

On 8 October, the Moon lies near Venus (see Special Events). The Moon is near Aldebaran on 22 October, and passes below Jupiter on 25 October. In the morning of 29 October, you'll find the crescent Moon near Regulus, with Mars to the left; the following morning, 30 October, the Moon is to the lower right of the Red Planet.

MOON		
Date	Time	Phase
5	1.35 am	New Moon
12	0.02 am	First Quarter
19	0.38 am	Full Moon
27	0.41 am	Last Quarter

SPECIAL EVENTS

On **8 October**, look out for the lovely sight of the crescent Moon above Venus, low in the south-west around 7.30 pm.

The Full Moon moves into the Earth's outer shadow on the night of **18/19 October**, in a penumbral eclipse (see April's Topic). The Moon's lower half should appear appreciably dimmer around mid-eclipse (0.50 am).

Debris from Halley's Comet smashes into Earth's atmosphere on **20/21 October**, causing the annual Orionid meteor shower. Unfortunately, all but the brightest shooting stars will be washed out this year by bright moonlight.

At 2 am on **27 October**, we see the end of British Summer Time for this year. Clocks go backwards by an hour.

▼ The Andromeda Galaxy, M31, photographed using a 200 mm telescope and CCD camera by Peter Shah from Meifod, Wales. Separate exposures of 50 minutes, 21 minutes, 27 minutes and 80 minutes were given through red, green, blue and H-alpha filters.

OCTOBER'S OBJECT

In the constellation of **Triangulum**, and just below the line of stars making up Andromeda, is the galaxy **M33**, which is also known as the **Triangulum Galaxy**. Although the Andromeda Galaxy is generally accepted to be the most distant object visible to the unaided eye, some amateur astronomers claim that, under exceptionally clear conditions, M33 can be seen. At magnitude +5.7 (with +6.5 being the naked-eye cut-off), this is plausible. But the light of the galaxy is spread out thinly, so there's very little contrast between the galaxy and the natural light of the night-time sky. With a distance of roughly 2.8 million light years, this makes M33 the furthest object visible to the naked eye.

The Triangulum Galaxy is a member of our Local Group. It's a spiral galaxy, rather more ragged and unbuttoned than the Milky Way and Andromeda. It's also a lot smaller – only half

> ***Viewing tip***
> Now that the nights are drawing in earlier, and becoming darker, it's a good time to pick out faint, fuzzy objects like M33, the Andromeda Galaxy and the Orion Nebula. But don't even think about it near the time of Full Moon – its light will drown them out. The best time to observe 'deep-sky objects' is just before or after New Moon. Check the Moon phases timetables in the book.

the diameter of the Milky Way, at 50,000 light years across. But it makes up for its size by containing one of the biggest star-forming regions known in the Universe. NGC 604 is a huge cloud of glowing gas and dust (small particles of cosmic soot) 1500 light years across – as compared to 24 light years for the Orion Nebula. It is already busy creating stars, and images from the Hubble Space Telescope have revealed over 200 young stars lurking in the mists that are between 15 and 60 times heavier than the Sun.

If you want to see M33 properly, you need a telescope. But don't use a high magnification – this will just spread the galaxy's light out, so that it's more difficult to see. Use a low magnification on a really transparent night.

OCTOBER'S PICTURE

At 2.5 million light years' distance, the **Andromeda Galaxy** is the closest spiral galaxy to our Milky Way. On a really clear, transparent night, it covers four moonwidths, but any lighting drowns its faint, outer spiral arms. Andromeda has two large satellite galaxies (NGC 205, right, and M32, left), which play gravitational tug-of war with their bigger neighbour. In 3–5 billion years' time, the Milky Way and Andromeda are set to collide, creating a vast elliptical galaxy – Milkomeda.

OCTOBER'S TOPIC
Asteroids

Asteroids are the 'salt of the Solar System' – primitive bodies that collided and coalesced together to create the planets we know today. Hundreds of thousands of these building blocks lie in a belt between the orbits of Mars and Jupiter, and, currently, NASA's spaceprobe DAWN is up close and personal with one of them – Vesta – in a mission to investigate the beginnings of our Solar System.

At 530 kilometres across, Vesta is one of the largest asteroids – and it's also the brightest. DAWN, now in orbit about the tiny, irregularly-shaped world, has sent back astonishing images of a heavily-cratered surface. Biggest of all is Rheasilvia (named after the mother of Romulus and Remus – a rumoured vestal virgin!). It's a colossal 460 kilometres wide across. The impact that created Rheasilvia is thought to have blasted out other asteroids, and even meteorites which have impacted the Earth.

This year, DAWN will move on, to encounter Ceres – the biggest asteroid – in 2015. Researchers believe that the two bodies have very different compositions, and that the mission should shed light on our own origins.

'A swarm of fireflies tangled in a silver braid' was the evocative description of the **Pleiades** star cluster by Alfred, Lord Tennyson, in his 1842 poem 'Locksley Hall'. Now rising in the east, the beautiful star cluster of the Seven Sisters is a sure sign that winter is on the way. From Greece to Australia, ancient myths independently describe the stars as a group of young girls being chased by an aggressive male – often **Aldebaran** or **Orion**. Polynesian navigators used the Pleiades to mark the start of their year. And farmers in the Andes rely on the visibility of the Pleiades as a guide to planting their potatoes: the brightness or faintness of the Seven Sisters depends on El Niño, which affects the forthcoming weather. Also this month: look out for Venus as a brilliant Evening Star, and a morning appearance of elusive Mercury.

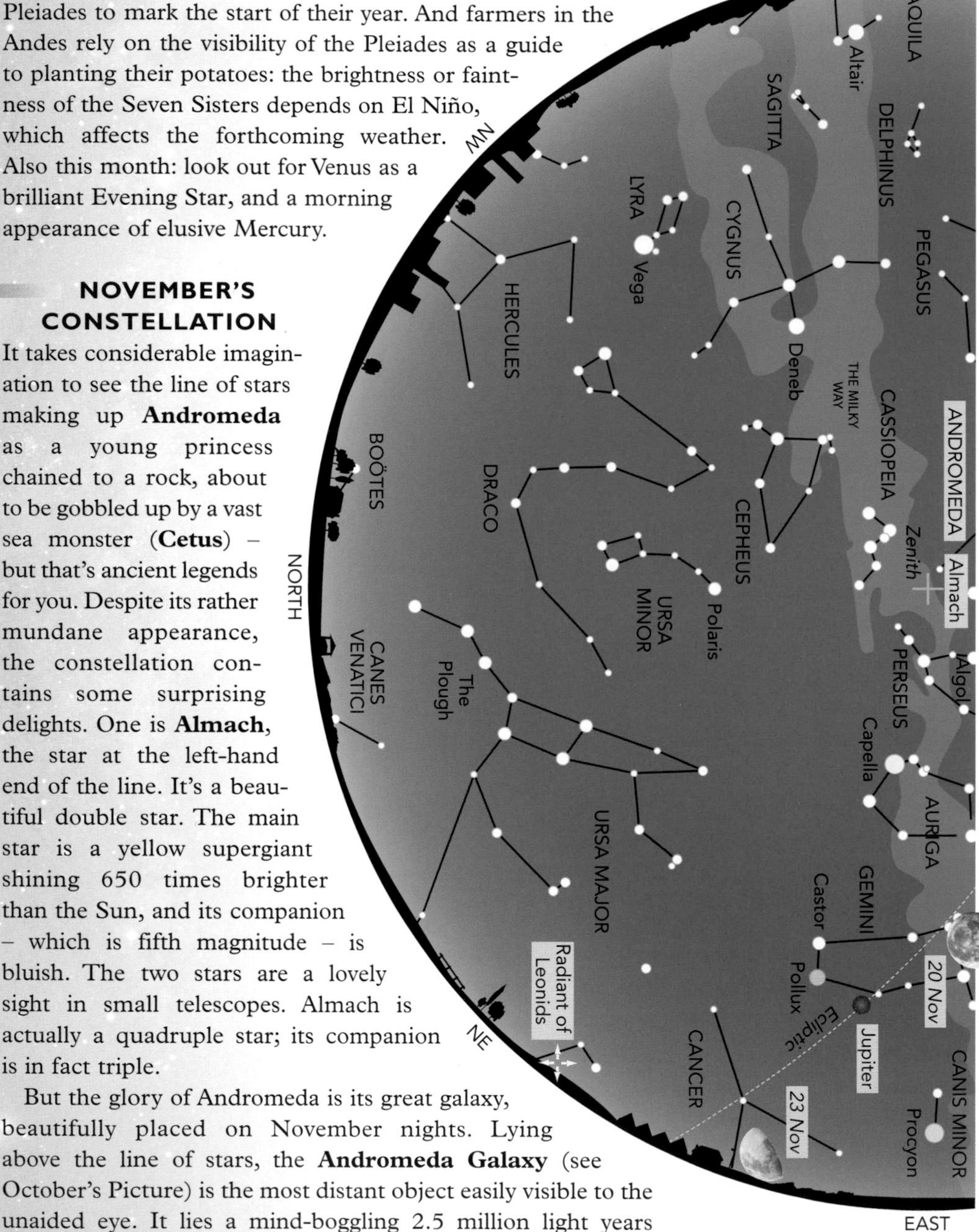

▼ *The sky at 10 pm in mid-November, with Moon positions at three-day intervals either side of Full Moon. The star positions are also correct for 11 pm at*

NOVEMBER'S CONSTELLATION

It takes considerable imagination to see the line of stars making up **Andromeda** as a young princess chained to a rock, about to be gobbled up by a vast sea monster (**Cetus**) – but that's ancient legends for you. Despite its rather mundane appearance, the constellation contains some surprising delights. One is **Almach**, the star at the left-hand end of the line. It's a beautiful double star. The main star is a yellow supergiant shining 650 times brighter than the Sun, and its companion – which is fifth magnitude – is bluish. The two stars are a lovely sight in small telescopes. Almach is actually a quadruple star; its companion is in fact triple.

But the glory of Andromeda is its great galaxy, beautifully placed on November nights. Lying above the line of stars, the **Andromeda Galaxy** (see October's Picture) is the most distant object easily visible to the unaided eye. It lies a mind-boggling 2.5 million light years

the beginning of November, and 9 pm at the end of the month. The planets move slightly relative to the stars during the month.

away. The Andromeda Galaxy is the biggest member of the Local Group – it's estimated to contain over 400 billion stars, and is a wonderful sight in binoculars or a small telescope.

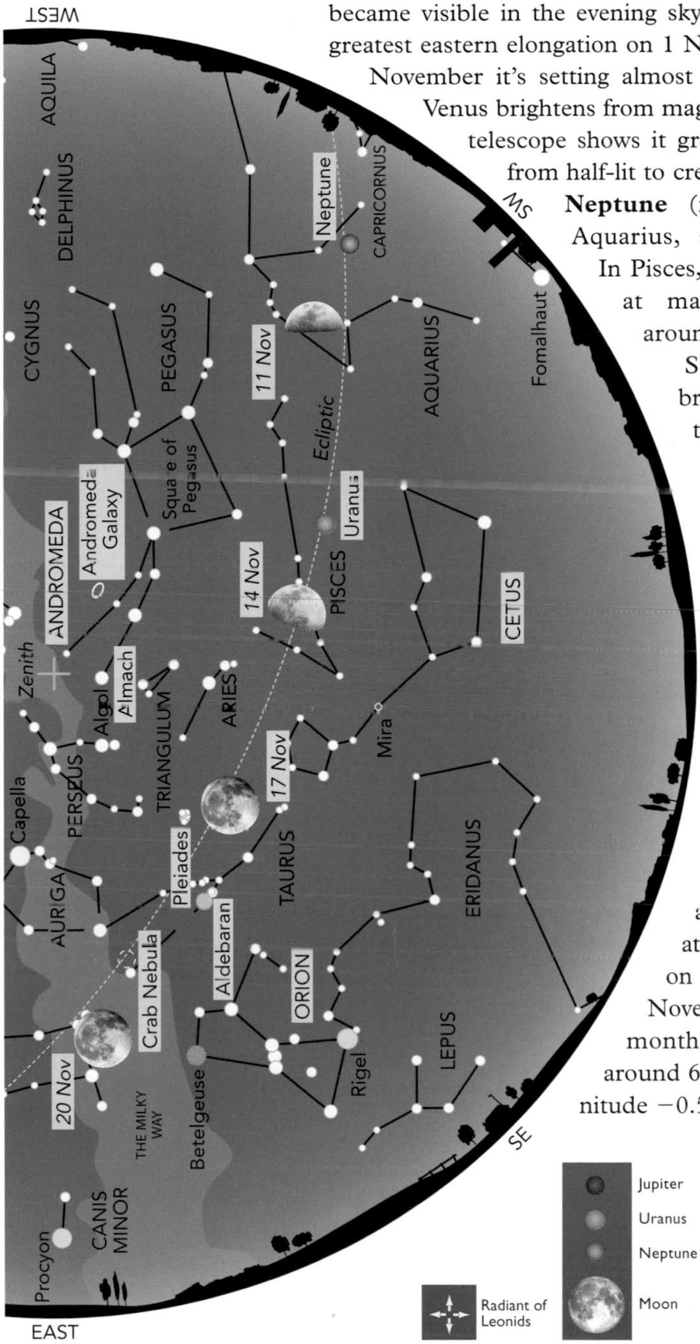

PLANETS ON VIEW

Venus roars into view this month, six months after it first became visible in the evening sky. The Evening Star is at its greatest eastern elongation on 1 November, and by the end of November it's setting almost three hours after the Sun. Venus brightens from magnitude −4.3 to −4.5, and a telescope shows it growing in size and changing from half-lit to crescent.

Neptune (magnitude +7.9) lies in Aquarius, setting about 11.30 pm. In Pisces, **Uranus** is slightly brighter at magnitude +5.8, and sets around 3 am.

Second only to Venus in brightness is **Jupiter** (magnitude −2.3), now rising about 7.30 pm in Gemini.

Mars, at magnitude +1.4, rises around 1 am. This month the Red Planet tracks under Leo and ends up in Virgo, passing less than an arcminute from sigma Leonis (magnitude +4.0) just before 4 am on 17 November.

If you're up before dawn, check out **Mercury** making its best morning appearance of the year: it's at greatest western elongation on 18 November. Between 14 November and the end of the month, look to the south-east around 6 am for a bright 'star' (magnitude −0.5), very close to the horizon.

MOON		
Date	Time	Phase
3	12.50 pm	New Moon
10	5.57 am	First Quarter
17	3.16 pm	Full Moon
25	7.28 pm	Last Quarter

Saturn is pulling out of the morning twilight, and passes only half a degree above Mercury on the morning of 26 November. At magnitude +0.8, the ringworld is four times fainter than Mercury.

MOON

You'll find the crescent Moon near Venus after sunset on 6 November (see Special Events); it lies above the Evening Star the following evening, 7 November. The Moon is close to Aldebaran on 18 November. On 21 and 22 November, the Moon lies near Jupiter. Regulus is the star next to the Moon on the night of 24/25 November. On the morning of 27 November, the Moon lies to the right of Mars.

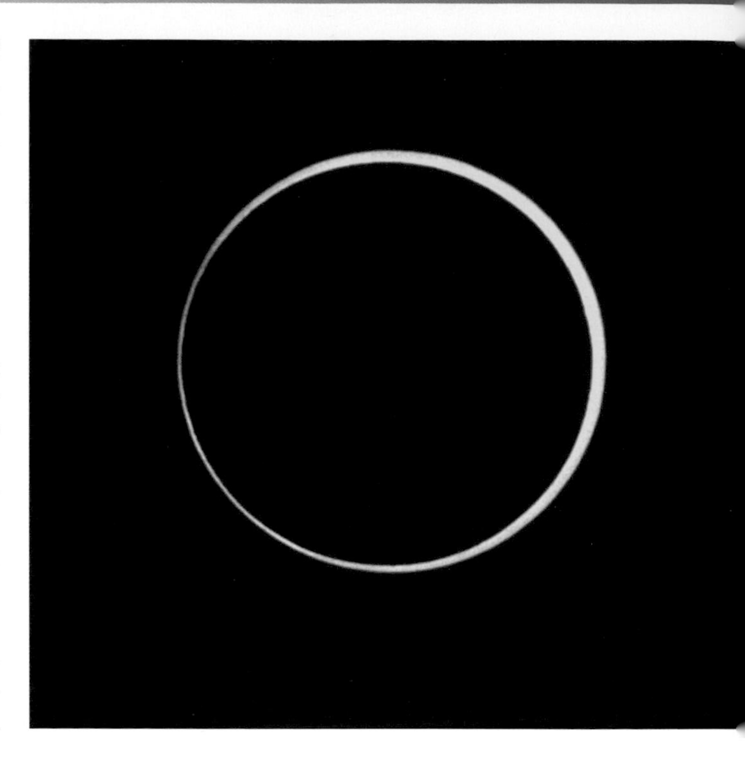

▲ *Annular eclipse of the Sun, photographed from Sumatra by Mike Harlow on 22 August 1998 using a 1000 mm f/10 lens with Mylar filter plus orange filter.*

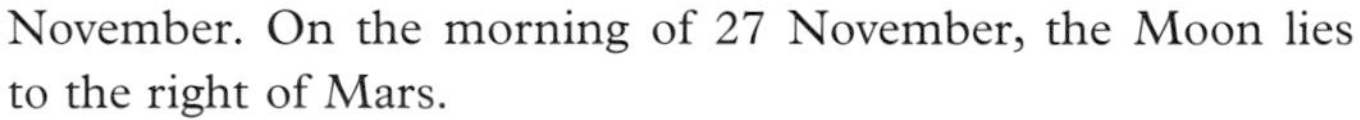

SPECIAL EVENTS

There's an unusual hybrid solar eclipse (see April's Topic) on **3 November**. It's annular at the beginning and end of its path – near Bermuda and Somalia, respectively – and total in the mid-Atlantic Ocean and equatorial Africa. Brazil, most of Africa and south Europe will witness a partial eclipse, but nothing is visible from the UK.

On **6 November**, the crescent Moon forms a striking tableau with Venus, low in the south-west around 6 pm.

The night of **17/18 November** sees the maximum of the **Leonid** meteor shower. But this year the shooting stars will be largely drowned out by the light of the Full Moon.

NOVEMBER'S OBJECT

Venus – the Planet of Love – is resplendent in our evening skies this month (see 'Viewing tip'). So brilliant and beautiful, she can even (it's said) cast a shadow in a really dark transparent sky. Her purity and lantern-like luminosity are beguiling – but looks are deceptive. Earth's twin in size, Venus could hardly be more different from our warm, wet world. The reason for the planet's brilliance is the highly reflective clouds that cloak its surface: probe under these palls of carbon dioxide and you find a planet out of hell. Volcanoes are to blame. They have created a runaway 'greenhouse effect' that has made Venus the hottest and most poisonous planet in the Solar System. At 460°C, this world is hotter than an oven. It boasts clouds of sulphuric acid. The pressure at its surface is around 90 Earth-atmospheres. So – if you visited Venus – you'd be simultaneously roasted, crushed, corroded and suffocated!

NOVEMBER'S PICTURE

An **annular eclipse of the Sun** – seen when the Moon is just too far away in its elliptical orbit around the Earth to cover the Sun entirely. Because there's a bright rim of light (or 'annulus') around the Sun, you don't get to see the glorious prominences or the solar corona (the Sun's outer atmosphere) during an annular eclipse. This month's eclipse will be partly annular, partly total, depending on your exact position on the Earth (see Special Events).

NOVEMBER'S TOPIC
Supernovae

With the star-wreck of the **Crab Nebula** sailing higher in the sky, it's a sobering reminder that some stars die young – and violently – with all the repercussions that will have on their planets and their lifeforms.

Supernovae – exploding stars – have been logged since ancient times. The Chinese called them 'guest stars'.

Stars more than eight times more massive than the Sun are doomed to an early death. They rip through the nuclear reactions that power their energy at a reckless rate. While our modest Sun converts hydrogen to helium in its core – making it shine – the biggies are far more ambitious. When the helium has run out, their gravity squeezes tighter, building successively new elements in the star's central nuclear reactor. All goes well – the star stays shining – until the core is made of iron.

Then it tries to fuse iron. It's a fatal mistake. Iron fusion takes in energy – and, as a result, the core catastrophically collapses. The star can't stand up to the shock. A burst of neutrinos blasts through its outer layers, blowing it apart.

At its maximum brightness, a supernova can outshine a whole galaxy of 100,000 million stars. And its aftermath is a tangled fireball of gases like the Crab Nebula.

The supernova hurls a cornucopia of elements into space – those from inside the dead star, and others created in the inferno of the explosion. So, in the end, a supernova is a phoenix: for out of its ashes, the seeds of life will arise.

Viewing tip

Venus is a real treat this month. If you have a small telescope, though, don't wait for the sky to get really dark. Seen against a black sky, the cloud-wreathed planet is so brilliant that it's difficult to make out anything on its disc. It's best to view the planet after the Sun has set, as soon as Venus is visible in the twilight glow, and you can then see the crescent-shaped planet appearing fainter against a pale blue sky.

This month sees the shortest day and the longest night. And the brilliant winter constellations are up there, in all their splendour. **Orion**, with his hunting dogs **Canis Major** and **Canis Minor**, is dominating the heavens, fighting his adversary **Taurus** (the Bull). To add to the excitement, the two dazzling planets Venus and Jupiter are adding a seasonal sparkle to the sky.

▼ The sky at 10 pm in mid-December, with Moon positions at three-day intervals either side of Full Moon. The star positions are also correct for 11 pm at

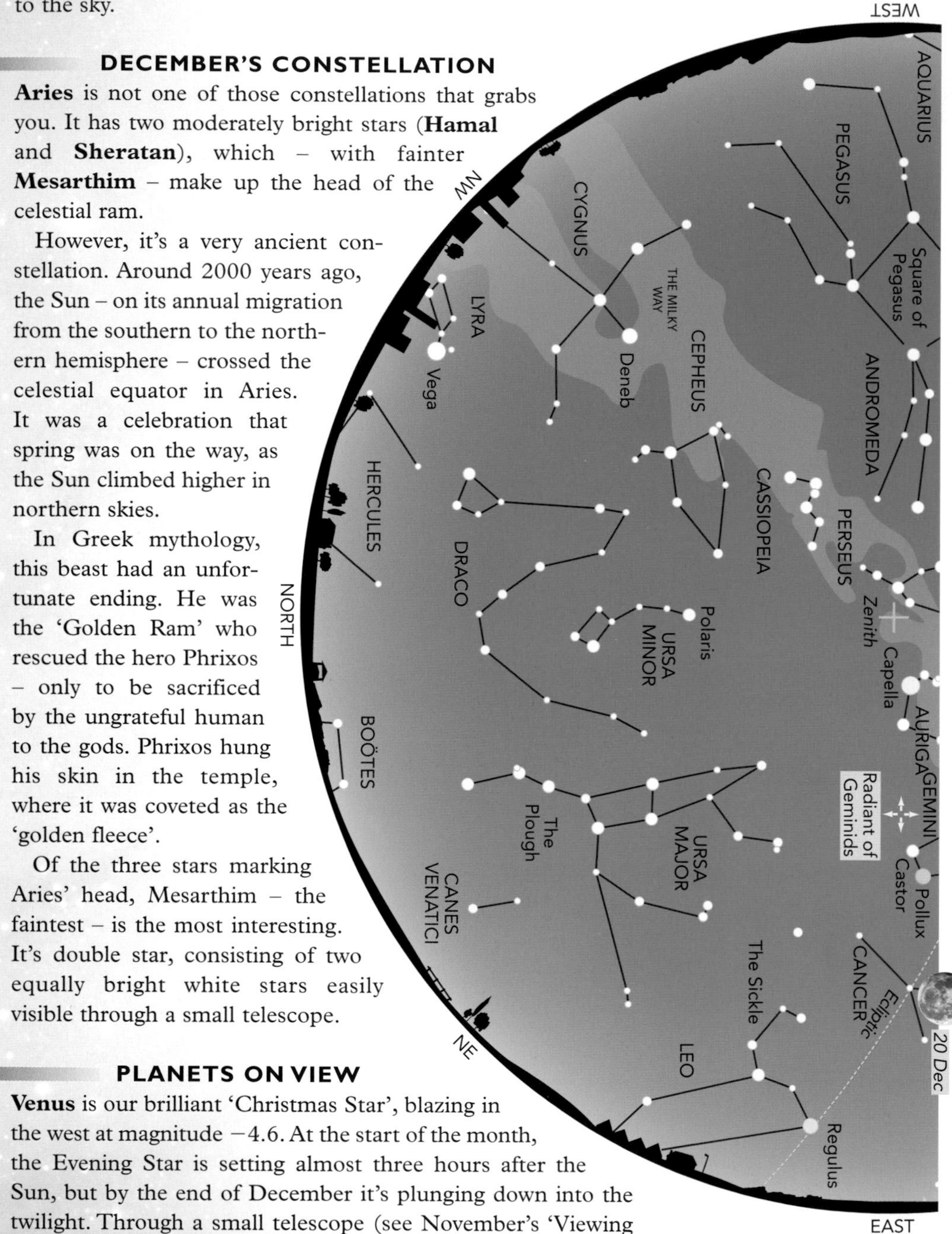

DECEMBER'S CONSTELLATION

Aries is not one of those constellations that grabs you. It has two moderately bright stars (**Hamal** and **Sheratan**), which – with fainter **Mesarthim** – make up the head of the celestial ram.

However, it's a very ancient constellation. Around 2000 years ago, the Sun – on its annual migration from the southern to the northern hemisphere – crossed the celestial equator in Aries. It was a celebration that spring was on the way, as the Sun climbed higher in northern skies.

In Greek mythology, this beast had an unfortunate ending. He was the 'Golden Ram' who rescued the hero Phrixos – only to be sacrificed by the ungrateful human to the gods. Phrixos hung his skin in the temple, where it was coveted as the 'golden fleece'.

Of the three stars marking Aries' head, Mesarthim – the faintest – is the most interesting. It's double star, consisting of two equally bright white stars easily visible through a small telescope.

PLANETS ON VIEW

Venus is our brilliant 'Christmas Star', blazing in the west at magnitude −4.6. At the start of the month, the Evening Star is setting almost three hours after the Sun, but by the end of December it's plunging down into the twilight. Through a small telescope (see November's 'Viewing

the beginning of December, and 9 pm at the end of the month. The planets move slightly relative to the stars during the month.

tip'), watch Venus grow larger as it approaches the Earth; at the same time, its shape changes into an ever-narrower crescent.

Over in Aquarius, **Neptune** (magnitude +7.9) is setting about 10 pm. And **Uranus**, in Pisces at magnitude +5.8, slips below the horizon around 1 am.

Mighty **Jupiter** is visible all night long, dominating Gemini with its magnitude −2.5 glare. On 10 December, the giant planet passes only a quarter of a degree from the star Wasat (magnitude +3.5).

Mars brightens during the month, from magnitude +1.2 to +0.9. The Red Planet rises at 0.30 am, and you'll find it in Virgo. On the night of 28/29 December, Mars is only half a degree from Porrima (magnitude +3.4).

Around 5 am, you'll find **Saturn** (magnitude +0.8) rising in the south-east, in Libra.

Mercury is lost in the Sun's glare this month.

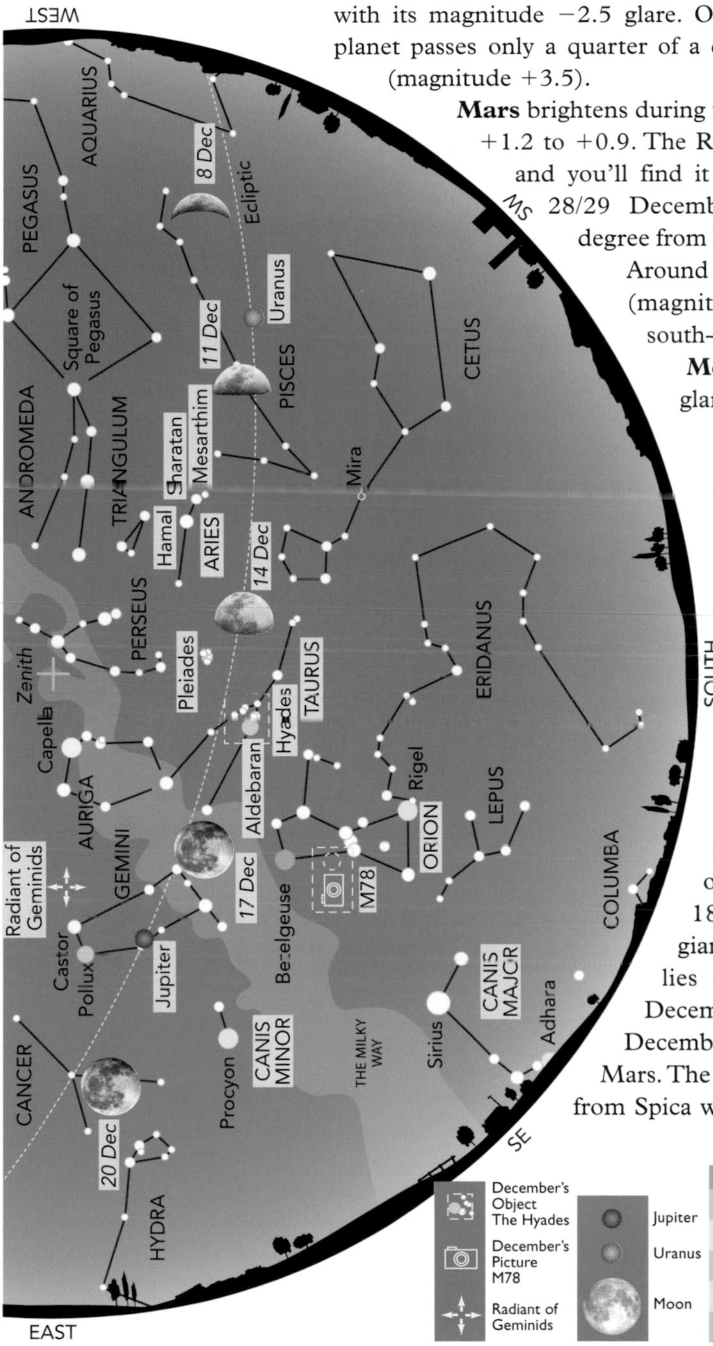

MOON

On the morning of 1 December, the narrow crescent Moon lies next to Saturn. The waxing crescent Moon passes Venus on 5 December (see Special Events). The Moon lies near Aldebaran on 15 December, as it passes through the fringes of the Hyades star cluster. The bright object near the Moon on 18 and 19 December is giant planet Jupiter. Regulus lies above the Moon on 22 December. The night of 25/26 December sees the Moon next to Mars. The Moon is only half a degree from Spica when it rises in the morning

MOON		
Date	Time	Phase
3	0.22 am	New Moon
9	3.12 pm	First Quarter
17	9.28 am	Full Moon
25	1.48 pm	Last Quarter

▲ *M78 in Orion, photographed by Peter Shah with a 200 mm telescope from Meifod, Wales, through red, green and blue filters plus H-alpha filter, with a total imaging time of 10½ hours.*

of 27 December. The crescent Moon lies just below Saturn before dawn on 29 December.

SPECIAL EVENTS

On the evening of **5 December**, the crescent Moon forms a lovely sight with Venus in the south-west around 6 pm.

The maximum of the **Geminid** meteor shower falls on **13/14 December**. These shooting stars are debris shed from an asteroid called Phaethon, and therefore quite substantial – and hence bright. This year we'll only see the most brilliant meteors, unfortunately, as the display is spoilt by bright moonshine.

The Winter Solstice occurs at 5.11 pm on **21 December**. As a result of the tilt of Earth's axis, the Sun reaches its lowest point in the heavens as seen from the northern hemisphere: we get the shortest days, and the longest nights.

DECEMBER'S OBJECT

The V-shaped **Hyades** star cluster, which forms the 'head' of **Taurus** (the Bull), doesn't hold a candle to the dazzling **Pleiades**. But it's the nearest star cluster to the Earth, and it forms the first rung of the ladder in establishing the cosmic distance scale. By measuring the motion of the stars in the

Viewing tip
This is the month when you may be thinking of buying a telescope as a Christmas present for a budding stargazer. Beware! Unscrupulous mail-order catalogues selling 'gadgets' often advertise small telescopes that boast huge magnifications. This is known as 'empty magnification' – blowing up an image that the lens or mirror simply doesn't have the ability to get to grips with, so all you see is a bigger blur. A rule of thumb is to use a maximum magnification no greater than twice the diameter of your lens or mirror in millimetres. So if you have a 100 mm reflecting telescope, go no higher than 200× magnification.

cluster, astronomers can establish their properties, and use these to find the distances to stars that are further away.

In legend, the Hyades feature in many myths – often as female figures (for example, the nymphs who cared for Bacchus as a baby). But the interpretations we like most are those of the Romans, who called the stars 'little pigs', and the Chinese vision of them as a 'rabbit net'.

Although **Aldebaran**, marking the bull's angry eye, looks as though it's part of the Hyades, this red giant just happens to lie in the same direction, and at less than half the distance.

The Hyades cluster lies 153 light years away (from the latest Hubble Space Telescope/Hipparcos satellite measurements), and contains about 200 stars. The stars are all around 625 million years old – very young on the stellar scale – and they could have a celestial twin. It turns out that Praesepe – the Beehive Cluster in Cancer – is the same age, and its stars are moving in the same direction. It may well be that the two clusters share a common birth.

DECEMBER'S PICTURE

Another of Orion's plentiful star-forming regions! This is **M78** – a 'reflection nebula'. The gas and dust are lit up by two particularly bright stars. The region is rich in cosmic dust (the dark swathes seen in this image) and gas – and is actively in the process of creating new stars. Nearly 50 T Tauri stars – young, unstable variable stars – have been detected in the nebula. M78 lies about 1600 light years away.

DECEMBER'S TOPIC
Origin of the Universe

Christmas is coming up, with all its associations with birth and beginnings. But how did our Universe begin? Luckily, we have some pretty firm evidence to answer that question.

Firstly, the Universe is expanding – on the largest scales, galaxies are moving apart from each other. If you 'rewind the tape', you'll find that the expansion started 13.7 billion years ago – a measurement that has only been tied down in recent years. Secondly, the Universe is not entirely cold – it's bathed in a radiation field of 2.7 degrees above Absolute Zero.

All these clues point to the origin of the Universe in a blisteringly hot 'Big Bang', which caused space to expand. The 'microwave background' of 2.7 degrees is the remnant of this birth in fire, cooled down by the relentless expansion to a mere shadow of its former self. Current observations show that the Universe is not just expanding, but accelerating – which means that it's destined to die by simply fading away.

There's always something to see in our Solar System, from planets to meteors or the Moon. These objects are very close to us – in astronomical terms – so their positions, shapes and sizes appear to change constantly. It is important to know when, where and how to look if you are to enjoy exploring Earth's neighbourhood. Here we give the best dates in 2013 for observing the planets and meteors (weather permitting!), and explain some of the concepts that will help you to get the most out of your observing.

Magnitudes

Astronomers measure the brightness of stars, planets and other celestial objects using a scale of *magnitudes*. Somewhat confusingly, fainter objects have higher magnitudes, while brighter objects have lower magnitudes; the most brilliant stars have negative magnitudes! Naked-eye stars range from magnitude −1.5 for the brightest star, Sirius, to +6.5 for the faintest stars you can see on a really dark night.

As a guide, here are the magnitudes of selected objects:

Sun	−26.7
Full Moon	−12.5
Venus (at its brightest)	−4.7
Sirius	−1.5
Betelgeuse	+0.4
Polaris (Pole Star)	+2.0
Faintest star visible to the naked eye	+6.5
Faintest star visible to the Hubble Space Telescope	+31

THE INFERIOR PLANETS

A planet with an orbit that lies closer to the Sun than the orbit of Earth is known as *inferior*. Mercury and Venus are the inferior planets. They show a full range of phases (like the Moon) from the thinnest crescents to full, depending on their position in relation to the Earth and the Sun. The diagram below shows the various positions of the inferior planets. They are invisible when at *conjunction*, when they are either behind the Sun, or between the Earth and the Sun, and lost in the latter's glare.

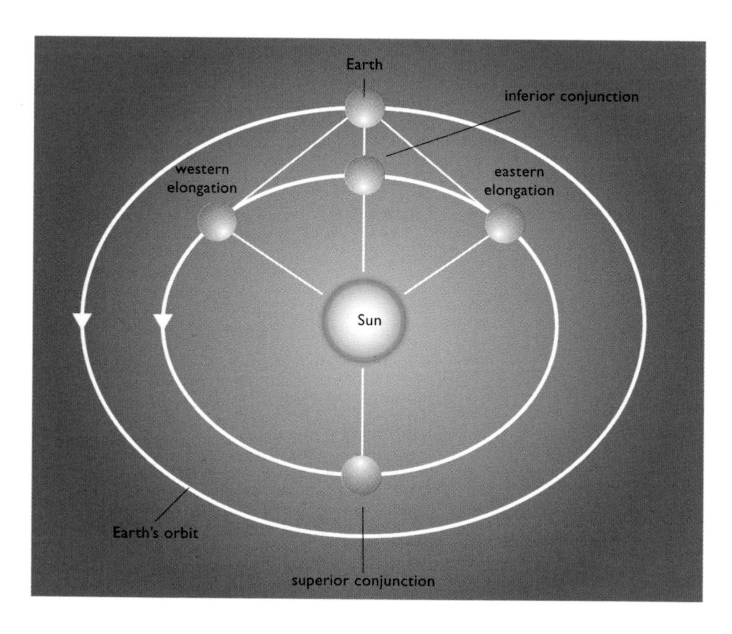

◀ *At eastern or western elongation, an inferior planet is at its maximum angular distance from the Sun. Conjunction occurs at two stages in the planet's orbit. Under certain circumstances, an inferior planet can transit across the Sun's disc at inferior conjunction.*

Mercury

Mercury's best evening appearance in 2013 occurs in February. Its subsequent evening performance in May–June is against a backdrop of bright twilight, while the October evening apparition is a complete wash-out as seen from Britain. In contrast, the morning appearances get better during the year. We're not going to see Mercury for its March apparition from the UK, but it's visible in the dawn skies of July–August. The planet's best morning appearance is in November.

Maximum elongations of Mercury in 2013

Date	Separation
16 February	18° east
31 March	28° west
12 June	24° east
30 July	20° west
9 October	25° east
18 November	20° west

Maximum elongation of Venus in 2013	
Date	Separation
1 November	47° east

Venus

Venus skulks in the twilight for almost all the year – as a Morning Star in January, and as an Evening Star from May onwards. Only in November and December does the brilliant planet live up to its reputation, as it rises from the dusk glow into the dark evening sky.

THE SUPERIOR PLANETS

The superior planets are those with orbits that lie beyond that of the Earth. They are Mars, Jupiter, Saturn, Uranus and Neptune. The best time to observe a superior planet is when the Earth lies between it and the Sun. At this point in a planet's orbit, it is said to be at *opposition*.

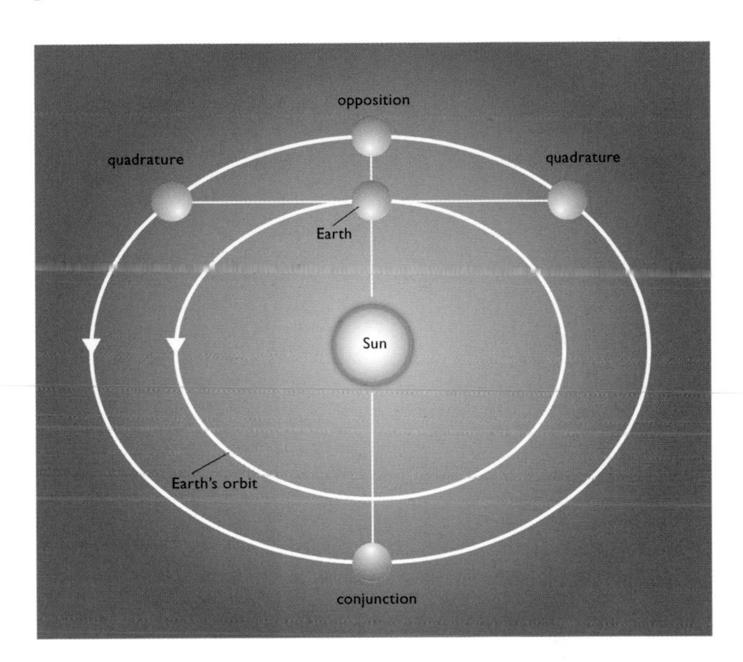

► *Superior planets are invisible at conjunction. At quadrature the planet is at right angles to the Sun as viewed from Earth. Opposition is the best time to observe a superior planet.*

Progress of Mars through the constellations	
January	Capricornus
Early July	Taurus
Late July – August	Gemini
September	Cancer
October – November	Leo
December	Virgo

Mars

The Red Planet does not have an opposition in 2013, and it puts on a generally poor show. You may just catch Mars low in the evening twilight in January, but it's then lost in the Sun's glare until July. From then until the end of the year, Mars is only visible after midnight.

Jupiter

The giant planet is brilliant in the evening sky from the beginning of the year until it sets in the twilight glow in May. Jupiter moves from Taurus into Gemini in June, and reappears in the morning sky in July. By the end of 2013, Jupiter is visible all night long. It doesn't have an opposition this year – we'll have to wait until January 2014!

Saturn

You'll find the ringed planet in Libra for most of the year, though it wanders into Virgo between May and August. Saturn is prominent in the morning sky at the start of 2013, and is visible all night when it reaches opposition on 28 April; it remains a feature of the evening sky until September. The ringworld reappears in the morning sky in November.

Uranus

Just perceptible to the naked eye, Uranus is visible in January and February, and then from July to December. It resides in Pisces all year, and is at opposition on 3 October.

Neptune

Lying in Aquarius all year, the most distant planet is at opposition on 27 August. Neptune can be seen – though only through a telescope – in January, and then from August to the end of the year.

Astronomical distances

For objects in the Solar System, such as the planets, we can give their distances from the Earth in kilometres. But the distances are just too huge once we reach out to the stars. Even the nearest star (Proxima Centauri) lies 25 million million kilometres away.

So astronomers use a larger unit – the *light year*. This is the distance that light travels in one year, and it equals 9.46 million million kilometres.

Here are the distances to some familiar astronomical objects, in light years:

Object	Light years
Proxima Centauri	4.2
Betelgeuse	640
Centre of the Milky Way	27,000
Andromeda Galaxy	2.5 million
Most distant galaxies seen by the Hubble Space Telescope	13 billion

SOLAR AND LUNAR ECLIPSES

Solar Eclipses

There are two solar eclipses in 2013. The first, on 10 May, is an annular solar eclipse, visible from northern Australia and the south Pacific; the whole of Australasia and most of the Pacific will witness a partial eclipse. The second, on 3 November, is an unusual hybrid solar eclipse: it's annular at the beginning and end of its path (near Bermuda and Somalia, respectively), and total in the mid-Atlantic Ocean and equatorial Africa. Brazil, most of Africa and south Europe will witness a partial eclipse.

Lunar Eclipses

There are three eclipses of the Moon in 2013: on 25 April, 25 May and 18/19 October. The April eclipse will be partial,

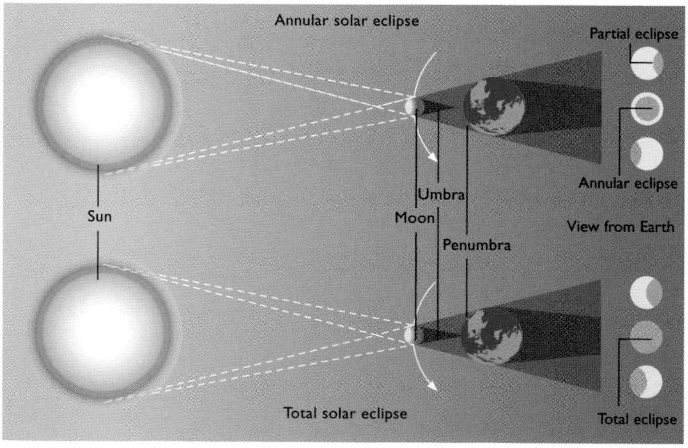

◀ *Where the dark central part (the umbra) of the Moon's shadow reaches the Earth, we see a total eclipse. People located within the penumbra see a partial eclipse. If the umbral shadow does not reach the Earth, we see an annular eclipse. This type of eclipse occurs when the Moon is at a distant point in its orbit and is not quite large enough to cover the whole of the Sun's disc.*

Dates of maximum for selected meteor showers	
Meteor shower	Date of maximum
Quadrantids	3/4 January
Lyrids	21/22 April
Eta Aquarids	5/6 May
Perseids	12/13 August
Orionids	20/21 October
Leonids	17/18 November
Geminids	13/14 December

whereas the eclipses in May and October are penumbral. Technically, the penumbral eclipse on 25 May is visible from the UK, but it will be hard to spot in the brightening dawn sky.

METEOR SHOWERS

Shooting stars – or *meteors* – are tiny particles of interplanetary dust, known as *meteoroids*, burning up in the Earth's atmosphere. At certain times of year, the Earth passes through a stream of these meteoroids (usually debris left behind by a comet) and we see a *meteor shower*. The point in the sky from which the meteors appear to emanate is known as the *radiant*. Most showers are known by the constellation in which the radiant is situated.

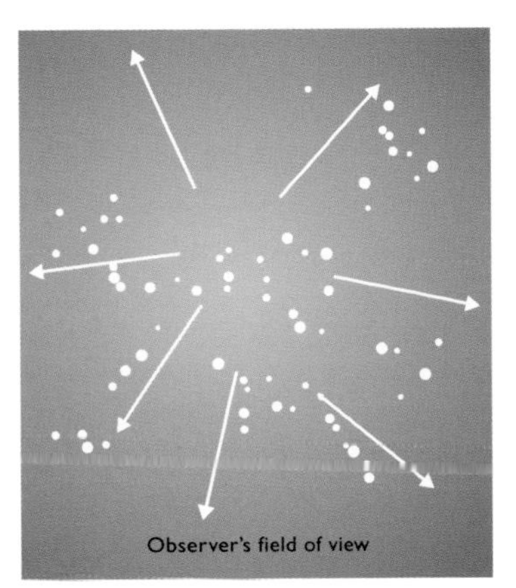

► *Meteors from a common source, occurring during a shower, enter the atmosphere along parallel trajectories. As a result of perspective, however, they appear to diverge from a single point in the sky – the radiant.*

When watching meteors for a co-ordinated meteor programme, observers generally note the time, seeing conditions, cloud cover, their own location, the time and brightness of each meteor, and whether it was from the main meteor stream. It is also worth noting details of persistent afterglows (trains) and fireballs, and making counts of how many meteors appear in a given period.

Angular separations

Astronomers measure the distance between objects, as we see them in the sky, by the angle between the objects in degrees (symbol °). From the horizon to the point above your head is 90 degrees. All around the horizon is 360 degrees.

You can use your hand, held at arm's length, as a rough guide to angular distances, as follows:

Width of index finger	1°
Width of clenched hand	10°
Thumb to little finger on outspread hand	20°

For smaller distances, astronomers divide the degree into 60 arcminutes (symbol ′), and the arcminute into 60 arcseconds (symbol ″).

COMETS

Comets are small bodies in orbit about the Sun. Consisting of frozen gases and dust, they are often known as 'dirty snowballs'. When their orbits bring them close to the Sun, the ices evaporate and dramatic tails of gas and dust can sometimes be seen.

A number of comets move round the Sun in fairly small, elliptical orbits in periods of a few years; others have much longer periods. Most really brilliant comets have orbital periods of several thousands or even millions of years. The exception is Comet Halley, a bright comet with a period of about 76 years. It was last seen with the naked eye in 1986.

Binoculars and wide-field telescopes provide the best views of comet tails. Larger telescopes with a high magnification are necessary to observe fine detail in the gaseous head (*coma*). Most comets are discovered with professional instruments, but a few are still found by experienced amateur astronomers.

None of the known comets is predicted to reach naked-eye brightness in 2013, but there's always a chance of a bright new comet putting in a surprise appearance.

Deep-sky objects are 'fuzzy patches' that lie outside the Solar System. They include star clusters, nebulae and galaxies. To observe the majority of deep-sky objects you will need binoculars or a telescope, but there are also some beautiful naked-eye objects, notably the Pleiades and the Orion Nebula.

The faintest object that an instrument can see is its *limiting magnitude*. The table gives a rough guide, for good seeing conditions, for a variety of small- to medium-sized telescopes.

We have provided a selection of recommended deep-sky targets, together with their magnitudes. Some are described in more detail in our monthly 'Object' features. Look on the appropriate month's map to find which constellations are on view, and then choose your objects using the list below. We have provided celestial coordinates for readers with detailed star maps. The suggested times of year for viewing are when the constellation is highest in the sky in the late evening.

Limiting magnitude for small to medium telescopes

Aperture (mm)	Limiting magnitude
50	+11.2
60	+11.6
70	+11.9
80	+12.2
100	+12.7
125	+13.2
150	+13.6

RECOMMENDED DEEP-SKY OBJECTS

Andromeda – autumn and early winter	
M31 (NGC 224) Andromeda Galaxy	3rd-magnitude spiral galaxy *RA 00h 42.7m Dec +41° 16'*
M32 (NGC 221)	8th-magnitude elliptical galaxy, a companion to M31 *RA 00h 42.7m Dec +40° 52'*
M110 (NGC 205)	8th-magnitude elliptical galaxy *RA 00h 40.4m Dec +41° 41'*
NGC 7662 Blue Snowball	8th-magnitude planetary nebula *RA 23h 25.9m Dec +42° 33'*
Aquarius – late autumn and early winter	
M2 (NGC 7089)	6th-magnitude globular cluster *RA 21h 33.5m Dec –00° 49'*
M72 (NGC 6981)	9th-magnitude globular cluster *RA 20h 53.5m Dec –12° 32'*
NGC 7293 Helix Nebula	7th-magnitude planetary nebula *RA 22h 29.6m Dec –20° 48'*
NGC 7009 Saturn Nebula	8th-magnitude planetary nebula *RA 21h 04.2m Dec –11° 22'*
Aries – early winter	
NGC 772	10th-magnitude spiral galaxy *RA 01h 59.3m Dec +19° 01'*
Auriga – winter	
M36 (NGC 1960)	6th-magnitude open cluster *RA 05h 36.1m Dec +34° 08'*
M37 (NGC 2099)	6th-magnitude open cluster *RA 05h 52.4m Dec +32° 33'*
M38 (NGC 1912)	6th-magnitude open cluster *RA 05h 28.7m Dec +35° 50'*
Cancer – late winter to early spring	
M44 (NGC 2632) Praesepe or Beehive	3rd-magnitude open cluster *RA 08h 40.1m Dec +19° 59'*
M67 (NGC 2682)	7th-magnitude open cluster *RA 08h 50.4m Dec +11° 49'*
Canes Venatici – visible all year	
M3 (NGC 5272)	6th-magnitude globular cluster *RA 13h 42.2m Dec +28° 23'*
M51 (NGC 5194/5) Whirlpool Galaxy	8th-magnitude spiral galaxy *RA 13h 29.9m Dec +47° 12'*
M63 (NGC 5055)	9th-magnitude spiral galaxy *RA 13h 15.8m Dec +42° 02'*
M94 (NGC 4736)	8th-magnitude spiral galaxy *RA 12h 50.9m Dec +41° 07'*
M106 (NGC4258)	8th-magnitude spiral galaxy *RA 12h 19.0m Dec +47° 18'*
Canis Major – late winter	
M41 (NGC 2287)	4th-magnitude open cluster *RA 06h 47.0m Dec –20° 44'*
Capricornus – late summer and early autumn	
M30 (NGC 7099)	7th-magnitude globular cluster *RA 21h 40.4m Dec –23° 11'*
Cassiopeia – visible all year	
M52 (NGC 7654)	6th-magnitude open cluster *RA 23h 24.2m Dec +61° 35'*
M103 (NGC 581)	7th-magnitude open cluster *RA 01h 33.2m Dec +60° 42'*
NGC 225	7th-magnitude open cluster *RA 00h 43.4m Dec +61 47'*
NGC 457	6th-magnitude open cluster *RA 01h 19.1m Dec +58° 20'*
NGC 663	Good binocular open cluster *RA 01h 46.0m Dec +61° 15'*
Cepheus – visible all year	
Delta Cephei	Variable star, varying between +3.5 and +4.4 with a period of 5.37 days. It has a magnitude +6.3 companion and they make an attractive pair for small telescopes or binoculars.
Cetus – late autumn	
Mira (omicron Ceti)	Irregular variable star with a period of roughly 330 days and a range between +2.0 and +10.1.
M77 (NGC 1068)	9th-magnitude spiral galaxy *RA 02h 42.7m Dec –00° 01'*

Coma Berenices – spring	
M53 (NGC 5024)	8th-magnitude globular cluster *RA 13h 12.9m Dec +18° 10'*
M64 (NGC 4286) Black Eye Galaxy	8th-magnitude spiral galaxy with a prominent dust lane that is visible in larger telescopes. *RA 12h 56.7m Dec +21° 41'*
M85 (NGC 4382)	9th-magnitude elliptical galaxy *RA 12h 25.4m Dec +18° 11'*
M88 (NGC 4501)	10th-magnitude spiral galaxy *RA 12h 32.0m Dec.+14° 25'*
M91 (NGC 4548)	10th-magnitude spiral galaxy *RA 12h 35.4m Dec +14° 30'*
M98 (NGC 4192)	10th-magnitude spiral galaxy *RA 12h 13.8m Dec +14° 54'*
M99 (NGC 4254)	10th-magnitude spiral galaxy *RA 12h 18.8m Dec +14° 25'*
M100 (NGC 4321)	9th-magnitude spiral galaxy *RA 12h 22.9m Dec +15° 49'*
NGC 4565	10th-magnitude spiral galaxy *RA 12h 36.3m Dec +25° 59'*
Cygnus – late summer and autumn	
Cygnus Rift	Dark cloud just south of Deneb that appears to split the Milky Way in two.
NGC 7000 North America Nebula	A bright nebula against the background of the Milky Way, visible with binoculars under dark skies. *RA 20h 58.8m Dec +44° 20'*
NGC 6992 Veil Nebula (part)	Supernova remnant, visible with binoculars under dark skies. *RA 20h 56.8m Dec +31 28'*
M29 (NGC 6913)	7th-magnitude open cluster *RA 20h 23.9m Dec +36° 32'*
M39 (NGC 7092)	Large 5th-magnitude open cluster *RA 21h 32.2m Dec +48° 26'*
NGC 6826 Blinking Planetary	9th-magnitude planetary nebula *RA 19 44.8m Dec +50° 31'*
Delphinus – late summer	
NGC 6934	9th-magnitude globular cluster *RA 20h 34.2m Dec +07° 24'*
Draco – midsummer	
NGC 6543	9th-magnitude planetary nebula *RA 17h 58.6m Dec +66° 38'*
Gemini – winter	
M35 (NGC 2168)	5th-magnitude open cluster *RA 06h 08.9m Dec +24° 20'*
NGC 2392 Eskimo Nebula	8–10th-magnitude planetary nebula *RA 07h 29.2m Dec +20° 55'*
Hercules – early summer	
M13 (NGC 6205)	6th-magnitude globular cluster *RA 16h 41.7m Dec +36° 28'*
M92 (NGC 6341)	6th-magnitude globular cluster *RA 17h 17.1m Dec +43° 08'*
NGC 6210	9th-magnitude planetary nebula *RA 16h 44.5m Dec +23 49'*
Hydra – early spring	
M48 (NGC 2548)	6th-magnitude open cluster *RA 08h 13.8m Dec –05° 48'*
M68 (NGC 4590)	8th-magnitude globular cluster *RA 12h 39.5m Dec –26° 45'*
M83 (NGC 5236)	8th-magnitude spiral galaxy *RA 13h 37.0m Dec –29° 52'*
NGC 3242 Ghost of Jupiter	9th-magnitude planetary nebula *RA 10h 24.8m Dec –18°38'*
Leo – spring	
M65 (NGC 3623)	9th-magnitude spiral galaxy *RA 11h 18.9m Dec +13° 05'*
M66 (NGC 3627)	9th-magnitude spiral galaxy *RA 11h 20.2m Dec +12° 59'*
M95 (NGC 3351)	10th-magnitude spiral galaxy *RA 10h 44.0m Dec +11° 42'*
M96 (NGC 3368)	9th-magnitude spiral galaxy *RA 10h 46.8m Dec +11° 49'*
M105 (NGC 3379)	9th-magnitude elliptical galaxy *RA 10h 47.8m Dec +12° 35'*
Lepus – winter	
M79 (NGC 1904)	8th-magnitude globular cluster *RA 05h 24.5m Dec –24° 33'*
Lyra – spring	
M56 (NGC 6779)	8th-magnitude globular cluster *RA 19h 16.6m Dec +30° 11'*
M57 (NGC 6720) Ring Nebula	9th-magnitude planetary nebula *RA 18h 53.6m Dec +33° 02'*
Monoceros – winter	
M50 (NGC 2323)	6th-magnitude open cluster *RA 07h 03.2m Dec –08° 20'*
NGC 2244	Open cluster surrounded by the faint Rosette Nebula, NGC 2237. Visible in binoculars. *RA 06h 32.4m Dec +04° 52'*
Ophiuchus – summer	
M9 (NGC 6333)	8th-magnitude globular cluster *RA 17h 19.2m Dec –18° 31'*
M10 (NGC 6254)	7th-magnitude globular cluster *RA 16h 57.1m Dec –04° 06'*
M12 (NCG 6218)	7th-magnitude globular cluster *RA 16h 47.2m Dec –01° 57'*
M14 (NGC 6402)	8th-magnitude globular cluster *RA 17h 37.6m Dec –03° 15'*
M19 (NGC 6273)	7th-magnitude globular cluster *RA 17h 02.6m Dec –26° 16'*
M62 (NGC 6266)	7th-magnitude globular cluster *RA 17h 01.2m Dec –30° 07'*
M107 (NGC 6171)	8th-magnitude globular cluster *RA 16h 32.5m Dec –13° 03'*
Orion – winter	
M42 (NGC 1976) Orion Nebula	4th-magnitude nebula *RA 05h 35.4m Dec –05° 27'*
M43 (NGC 1982)	5th-magnitude nebula *RA 05h 35.6m Dec –05° 16'*
M78 (NGC 2068)	8th-magnitude nebula *RA 05h 46.7m Dec +00° 03'*
Pegasus – autumn	
M15 (NGC 7078)	6th-magnitude globular cluster *RA 21h 30.0m Dec +12° 10'*
Perseus – autumn to winter	
M34 (NGC 1039)	5th-magnitude open cluster *RA 02h 42.0m Dec +42° 47'*
M76 (NGC 650/1) Little Dumbbell	11th-magnitude planetary nebula *RA 01h 42.4m Dec +51° 34'*

NGC 869/884 Double Cluster	Pair of open star clusters *RA 02h 19.0m Dec +57° 09'* *RA 02h 22.4m Dec +57° 07'*
Pisces – autumn	
M74 (NGC 628)	9th-magnitude spiral galaxy *RA 01h 36.7m Dec +15° 47'*
Puppis – late winter	
M46 (NGC 2437)	6th-magnitude open cluster *RA 07h 41.8m Dec –14° 49'*
M47 (NGC 2422)	4th-magnitude open cluster *RA 07h 36.6m Dec –14° 30'*
M93 (NGC 2447)	6th-magnitude open cluster *RA 07h 44.6m Dec –23° 52'*
Sagitta – late summer	
M71 (NGC 6838)	8th-magnitude globular cluster *RA 19h 53.8m Dec +18° 47'*
Sagittarius – summer	
M8 (NGC 6523) Lagoon Nebula	6th-magnitude nebula *RA 18h 03.8m Dec –24° 23'*
M17 (NGC 6618) Omega Nebula	6th-magnitude nebula *RA 18h 20.8m Dec –16° 11'*
M18 (NGC 6613)	7th-magnitude open cluster *RA 18h 19.9m Dec –17 08'*
M20 (NGC 6514) Trifid Nebula	9th-magnitude nebula *RA 18h 02.3m Dec –23° 02'*
M21 (NGC 6531)	6th-magnitude open cluster *RA 18h 04.6m Dec –22° 30'*
M22 (NGC 6656)	5th-magnitude globular cluster *RA 18h 36.4m Dec –23° 54'*
M23 (NGC 6494)	5th-magnitude open cluster *RA 17h 56.8m Dec –19° 01'*
M24 (NGC 6603)	5th-magnitude open cluster *RA 18h 16.9m Dec –18° 29'*
M25 (IC 4725)	5th-magnitude open cluster *RA 18h 31.6m Dec –19° 15'*
M28 (NGC 6626)	7th-magnitude globular cluster *RA 18h 24.5m Dec –24° 52'*
M54 (NGC 6715)	8th-magnitude globular cluster *RA 18h 55.1m Dec –30° 29'*
M55 (NGC 6809)	7th-magnitude globular cluster *RA 19h 40.0m Dec –30° 58'*
M69 (NGC 6637)	8th-magnitude globular cluster *RA 18h 31.4m Dec –32° 21'*
M70 (NGC 6681)	8th-magnitude globular cluster *RA 18h 43.2m Dec –32° 18'*
M75 (NGC 6864)	9th-magnitude globular cluster *RA 20h 06.1m Dec –21° 55'*
Scorpius (northern part) – midsummer	
M4 (NGC 6121)	6th-magnitude globular cluster *RA 16h 23.6m Dec –26° 32'*
M7 (NGC 6475)	3rd-magnitude open cluster *RA 17h 53.9m Dec –34° 49'*
M80 (NGC 6093)	7th-magnitude globular cluster *RA 16h 17.0m Dec –22° 59'*
Scutum – mid to late summer	
M11 (NGC 6705) Wild Duck Cluster	6th-magnitude open cluster *RA 18h 51.1m Dec –06° 16'*
M26 (NGC 6694)	8th-magnitude open cluster *RA 18h 45.2m Dec –09° 24'*
Serpens – summer	
M5 (NGC 5904)	6th-magnitude globular cluster *RA 15h 18.6m Dec +02° 05'*
M16 (NGC 6611)	6th-magnitude open cluster, surrounded by the Eagle Nebula. *RA 18h 18.8m Dec –13° 47'*
Taurus – winter	
M1 (NGC 1952) Crab Nebula	8th-magnitude supernova remnant *RA 05h 34.5m Dec +22° 00'*
M45 Pleiades	1st-magnitude open cluster, an excellent binocular object. *RA 03h 47.0m Dec +24° 07'*
Triangulum – autumn	
M33 (NGC 598)	6th-magnitude spiral galaxy *RA 01h 33.9m Dec +30° 39'*
Ursa Major – all year	
M81 (NGC 3031)	7th-magnitude spiral galaxy *RA 09h 55.6m Dec +69° 04'*
M82 (NGC 3034)	8th-magnitude starburst galaxy *RA 09h 55.8m Dec +69° 41'*
M97 (NGC 3587) Owl Nebula	12th-magnitude planetary nebula *RA 11h 14.8m Dec +55° 01'*
M101 (NGC 5457)	8th-magnitude spiral galaxy *RA 14h 03.2m Dec +54° 21'*
M108 (NGC 3556)	10th-magnitude spiral galaxy *RA 11h 11.5m Dec +55° 40'*
M109 (NGC 3992)	10th-magnitude spiral galaxy *RA 11h 57.6m Dec +53° 23'*
Virgo – spring	
M49 (NGC 4472)	8th-magnitude elliptical galaxy *RA 12h 29.8m Dec +08° 00'*
M58 (NGC 4579)	10th-magnitude spiral galaxy *RA 12h 37.7m Dec +11° 49'*
M59 (NGC 4621)	10th-magnitude elliptical galaxy *RA 12h 42.0m Dec +11° 39'*
M60 (NGC 4649)	9th-magnitude elliptical galaxy *RA 12h 43.7m Dec +11° 33'*
M61 (NGC 4303)	10th-magnitude spiral galaxy *RA 12h 21.9m Dec +04° 28'*
M84 (NGC 4374)	9th-magnitude elliptical galaxy *RA 12h 25.1m Dec +12° 53'*
M86 (NGC 4406)	9th-magnitude elliptical galaxy *RA 12h 26.2m Dec +12° 57'*
M87 (NGC 4486)	9th-magnitude elliptical galaxy *RA 12h 30.8m Dec +12° 24'*
M89 (NGC 4552)	10th-magnitude elliptical galaxy *RA 12h 35.7m Dec +12° 33'*
M90 (NGC 4569)	9th-magnitude spiral galaxy *RA 12h 36.8m Dec +13° 10'*
M104 (NGC 4594) Sombrero Galaxy	Almost edge-on 8th-magnitude spiral galaxy. *RA 12h 40.0m Dec –11° 37'*
Vulpecula – late summer and autumn	
M27 (NGC 6853) Dumbbell Nebula	8th-magnitude planetary nebula *RA 19h 59.6m Dec +22° 43'*

FINDING YOUR WAY

One challenge that every budding astronomer has to confront is that of finding their way around the sky. The star maps in *Stargazing 2013* are a great way to begin and, of course, they also include the planets. But there are plenty of reasons why you might need to go a bit deeper, or to work out something else. What time is sunset? What's that bright star I saw at 3 am? Or maybe if you are more advanced in astronomy you will want to know how to locate the Crab Nebula, or what stars will be visible on your Mediterranean holiday. This article looks at the various options for getting to know the sky better.

PHILIP'S PLANISPHERE

For many years, a planisphere has been the first step for anyone who wants to get to grips with the sky. To be grand, it is an analog computer – but in practical terms it is a map of the whole sky with a movable overlay which you can position according to the particular date and time when you will be observing. The overlay has an oval window through which you can see a map of the sky for that moment, and indeed all the other dates and times which match up for that particular setting. The window shows the whole sky visible at the time, just like those in *Stargazing 2013*, but it is oval rather than circular because of the difficulty of representing a large curved sphere on a flat surface. There is some distortion of constellations in the far south of the sky as a result, but they are still easy to pick out.

The dates around the edge can also be used to work out the position of the Sun in the sky. It moves steadily along the ecliptic in line with these dates, so you simply put a straight edge from the middle of the planisphere to the date at the edge, and notice where this crosses the ecliptic. Having found the Sun's position against the stars, you now know when it is in the sky on that particular date, so when it is on either the eastern or western horizon you can estimate with fair accuracy the time of sunrise or sunset.

A single map cannot show the position of the Moon or planets, but in the case of the bright planets Venus, Mars, Jupiter and Saturn there is a table of their positions for each month for many years to come, printed on the back of the planisphere. As with the Sun, you simply align a straight edge between the position shown at the edge of the planisphere and the middle, and you can see where the planet will be.

There are separate planispheres for a range of latitudes. As well as 51.5°N, suitable for the whole of the British Isles, there are 42°N, 32°N, 23.5°N and 35°S, each model being usable for several degrees on either side of the stated latitude.

▼ *A Philip's Planisphere for latitude 51.5°N. Lining up the date and time using the scales around the edge shows the whole sky for that time and date.*

STAR ATLASES

The advantage of a planisphere is that it costs and weighs very little, can easily be packed in a suitcase, is instantly ready for use, and of course needs no batteries. But there comes a point where you might need a more advanced sky map. There are of course many star atlases to choose from, and modesty does not fully prevent me from mentioning my own *Night Sky Atlas*, published by Philip's, with maps by Wil Tirion. This includes much information on observing the sky and objects of interest in each constellation that are visible with the naked eye, binoculars and small telescopes. It shows stars down to a practical naked-eye limit of magnitude 5.5. Deeper coverage is provided by *Norton's Star Atlas and Reference Handbook*, published by Addison Wesley, which goes to magnitude 6.5, and *Cambridge Star Atlas*, from CUP, which goes down to about magnitude 7. Both also use maps by Dutch cartographer Wil Tirion.

▲ *The free* Stellarium *software shows a realistic view of the sky which you can set for any date and time. You can also zoom in to see fainter stars.*

COMPUTER SOFTWARE

At one time, anyone who needed a map showing stars fainter than the naked-eye limit would buy a printed atlas, but these are now rarely used. In their place are computer-based star map or planetarium programs. These use star catalogues which show stars down to about magnitude 15, combined with other catalogues of deep-sky and other objects, and routines that calculate the positions of planets and other Solar System bodies. So in one package you can see stars, planets, asteroids and deep-sky objects, and print them out if required at any scale. They are not limited to the here-and-now, but can be used for any chosen date, so you can plan your observing. You can even see what the sky would have looked like at some date in the past. Slow changes in the direction of Earth's axis, known as precession, mean that over hundreds of years the stars visible from a specific location do change slightly. In Roman times, for example, what we now call the Pole Star was many degrees from where it is now. And if you go back to the time

▼ *A section from Chris Marriott's* SkyMap Pro 11, *showing the region of the Orion Nebula and stars down to magnitude 11.5.*

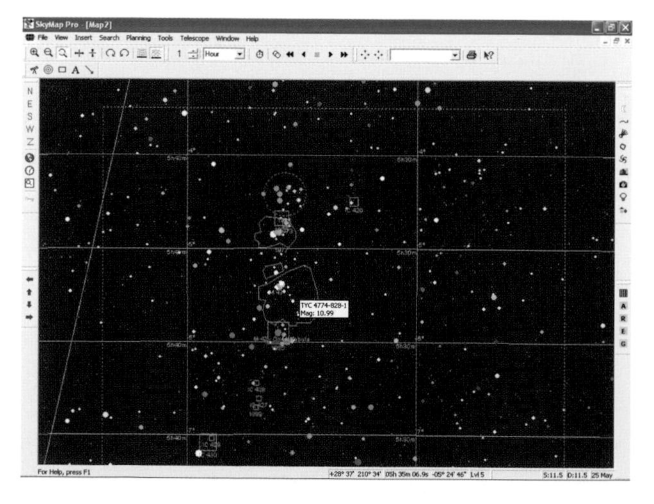

▲ *Programs such as* Stellarium *and* SkyMap *can be used to control a telescope – just click to slew it to a chosen object. Here, a digital SLR and autoguider are also being controlled from indoors.*

when Stonehenge was still under construction, its builders could have seen the Southern Cross – now only visible from the Tropics or farther south.

Though the stars themselves don't move appreciably from year to year, precession means that star positions get out of date, so printed maps have to be redrawn from time to time. This makes little difference for casual stargazing, but for finding ephemeral objects such as asteroids or comets all positions have to be on the same basis. A computer program takes all this into account and will allow you to plot the paths of newly discovered objects from their orbital details, available online. To add to their value, the programs can often be linked to many telescope mountings so that they not only show precisely where the telescope is pointing, but can also be used to direct the instrument to a chosen object. When everything is working well, you can simply call up an object from the program's database and the telescope will slew to the object that you want to observe. This makes possible remote observing using an imaging device on the telescope, with the 'observer' keeping warm indoors and controlling everything from the computer.

There are proprietary programs that will do all this, but you can get free software as well. One of the most popular is *Stellarium*, which is a free download. One of its merits is that it presents a very realistic view of the sky, in which you can change the amount of haze and even the amount of star twinkling.

Anyone with *Google Earth* on their computer – a free download that allows you to explore the Earth's surface in great detail – has access to *Google Sky*. Click on the sky symbol when in *Google Earth* and the application zooms into a black sky to show the stars overhead at the time. With 'Backyard Astronomy' selected in the menu on the left of the screen, constellation patterns are shown and you can zoom into stars and deep-sky objects. The planets are shown by symbols, and you can be taken to a planet by clicking on its name in the left-hand pane. However, the system lacks the functionality of a planetarium program which realistically shows you the sky as you would see it from your location, crucially with a horizon to make it clear which objects are actually visible at any time. The sky data is a mosaic of sky survey photos, and is very patchy with many strange artefacts. These have led to many discussions about UFOs or areas that are blanked out and which we are apparently not being allowed to see.

WEBSITES

For a comprehensive list of astronomical sky software, go to:

http://astro.nineplanets.org/astrosoftware.html

Site for the latest orbital data on comets and other Solar System bodies:

http://ssd.jpl.nasa.gov/sbdb.cgi#top

Useful information on transient phenomena available from:

www.heavens-above.com

APPS

These days, apps are all the rage. This refers to applications for either advanced mobile phones (smartphones) or for tablet computers, of which the best known is the iPad. These provide mobile access to many computing tasks in a thin, lightweight device about the size of a tablemat. For each device there are thousands of apps, many of which are connected with astronomy, but the most relevant are those which will display a sky map at the touch of a button, and, most usefully, react to the tilt of the device so that as you hold it up to the sky, it displays the stars and planets visible in that direction. Even on a cloudy day you can hold them up to the sky, and they give what appears to be a window into the sky, showing the view that you would get if you could see the stars at that moment. Hold it in the direction of the Sun, and there it is, with stars behind it. For all the advantages of other methods of finding what's in the sky, these have the most immediate appeal. You see a star in the sky and simply hold the device up to see it labelled, with the option in many cases of finding out more about the objects being shown. The ready availability of these apps will help many people become more interested in astronomy.

To use such an app properly, however, your device must be equipped with GPS (global positioning system, which gives your precise location), and an electronic compass and other sensors that tell it which way it is being tilted. Only the more expensive devices in each range have these. There is a range of software available, which depends on the device you have. Many mobile phones have the Android operating system, for which *Google SkyMap* is available free of charge. This is rather different from *Google Sky*, and does display your horizon and a recognizable sky. Users of Apple iPhones and iPads have different apps available.

Users do comment that sometimes they get odd results from these apps, with the display failing to move as it should or to display the correct stars, but they rate highly among favourite apps, even among non-astronomers. To add to their appeal for astronomers, cables are available for some devices that will even allow them to control a suitably equipped telescope in the same way as the computer planetarium programs. For some years there have been purpose-built devices which will name any astronomical object at which they are pointed, such as the Celestron SkyScout. However, these cost as much as, or even more, than a more versatile smartphone.

▼ *A Samsung Galaxy smartphone with* Google SkyMap *showing a map of the sky in the direction it is being held – here, Orion.*